Computing System Reliability

Models and Analysis

Computing System Reliability

Models and Analysis

Min Xie

Yuan-Shun Dai

and

Kim-Leng Poh

National University of Singapore
Singapore

Springer Science+Business Media, LLC

ISBN 978-1-4757-8804-4 ISBN 978-0-306-48636-4 (eBook)
DOI 10.1007/978-0-306-48636-4

Originally published by Kluwer Academic/Plenum Publishers, New York in 2004
Softcover reprint of the hardcover 1st edition 2004

http://www.wkap.nl/

10 9 8 7 6 5 4 3 2 1

A C.I.P. record for this book is available from the Library of Congress

Preface

Computing systems are widely used today and in many areas they serve the key function in achieving highly complicated and safety-critical mission. At the same time, the size and complexity of computing systems have continued to increase, making its performance evaluation more difficult than ever before.

The purpose of this book is to provide a comprehensive coverage of tools and techniques for computing system reliability modeling and analysis. Reliability analysis is a useful tool in evaluating the performance of complex systems. Intensive studies have been carried out to improve the likelihood for computing systems to perform satisfactorily in operation.

Software and hardware are two major building blocks in computing systems. They have to work together successfully to complete many critical computing tasks. This book systematically studies the reliability of software, hardware and integrated software/hardware systems. It also introduces typical models in the reliability analysis of the distributed/networked systems, and then further develops some new models and analytical tools.

"Grid" computing system has emerged as an important new field, distinguished from conventional distributed computing systems by its focus on large-scale resource sharing, innovative applications, and, in many cases, high-performance orientation. This book also presents general reliability models for the grid and discusses analytical tools to estimate the grid reliability related to the resource management system, wide-area network communication, and parallel running programs with multiple shared resources.

Furthermore, this book introduces the basic reliability theories and models for various multi-state systems. Based on the models, some interesting decision problems in system design and resource allocation are further discussed.

This book is organized as follows.

Chapter 1 provides an introduction to the field of computing systems and reliability analysis. Simple reliability concepts are also discussed. Chapter 2 provides the basic knowledge in reliability analysis and summarizes some common techniques for analyzing the computing system reliability. The fundamentals of Markov processes and Nonhomogeneous Poisson processes (NHPP) are also introduced, which are essential tools used in this book.

Chapters 3 and 4 present important models for the reliability analysis of hardware and software systems, respectively. They are useful when hardware and software issues are dealt with separately at the system analysis stage. Chapter 5 discusses the models for integrated systems. This is essential in computing system analysis as both software and hardware systems have to work together.

In Chapter 6, the reliability of various distributed computing systems which incorporate the network communication into the hardware/software reliability is studied. The distributed computing system is a common and widely-used networked system and hence a chapter is devoted to this.

The reliability of grid computing systems, which is a new direction in computing technology, is studied in Chapter 7. Since the grid reliability is difficult to evaluate due to its wide-area, heterogeneous and time various characteristics, we initially construct the reliability models for the different parts of the grid, including resource management system, large-scale network, distributed software and resources.

Finally, Chapter 8 studies the multi-state system reliability. Some optimization models in the system design and resource allocation are presented in Chapter 9. This is an area where research is going on and further development is needed.

The basic chapters in this book are Chapters 3-7. Readers familiar with basic reliability can start from Chapter 3 directly. Chapters 8 and 9 are on advanced topics and can be read by those interested in those specific topics.

Many models and results found in the literature and from our research are presented in the book. It is hoped that these approaches are easily implemented by practitioners as well. In addition, many examples are accompanied with those approaches.

The book serves as reference book for students, professors, engineers and researchers in related science and engineering field. It can be used for graduate and senior undergraduate courses. Researchers and students should find many ideas useful in their academic work.

The readers should have some basic knowledge in probability and calculus. However, difficult details are omitted to benefit the general audience. References are given so that further details can be found for those who are interested in more specific results.

M. Xie
Y. S. Dai
K. L. Poh

Acknowledgements

This book has evolved over the past ten years. We would like to thank our collaborators and students who have worked on one or more topics covered in the book. Especially, G. Y. Hong, B. Yang, S. L. Ho and G. Q. Liu have contributed a significant amount to the work presented in the book.

We are fortunate to have worked closely with many overseas colleagues, such as L. R. Cui, O. Gaudoin, P. K. Kapur, C. D. Lai, G. Levitin, D. N. P. Murthy, K. Way, C. Wohlin, M. Zhao, among others. This has helped broaden our view which is needed for a book as such. Many people worldwide have been interested in our work and we are grateful to T. Dohi, K. Kanoun, T. Khoshgoftaar, S. Y. Kuo, M. R. Lyu, J. Musa, S. Osaki, H. Pham, N. Schneidewind, Y. Tohma, K. S. Trivedi, M. Vouk, L. Walls, S. Yamada, M. J. Zuo and others for their help in our research.

Our research is supported by the National University of Singapore. We are also grateful to all staff and other students in the Department of Industrial and Systems Engineering for their help in one way or another.

The effort of Gary Folven of Kluwer and other staff at the Kluwer Academic Publisher is also appreciated. The idea of putting together this book was firmed up after his trip to Singapore at the beginning of the millennium.

Finally, we would like to thank our families for their understanding and support in all these years.

Contents

CHAPTER 1

INTRODUCTION

1.1. Need for Computing System Reliability Analysis

Computing has been the fastest developing technology during the last century. Computing systems are widely used in many areas, and they are desired to achieve various complex and safety-critical missions. The applications of the computing systems have now crossed many different fields and can be found in different products, for example, air traffic control systems, nuclear power plants, aircrafts, real-time military systems, telephone switching, bank auto-payment, hospital patient monitoring systems, and so forth.

The size and complexity of the computing systems has increased from one single processor to multiple distributed processors, from individual-separated systems to networked-integrated systems, from small-scale program running to large-scale resource sharing, and from local-area computation to global-area collaboration. A computing system today may contain many processors and communication channels and it may cover a wide area all over the world. They combine both software and hardware that have to function together to complete various tasks. They may incorporate multiple states and their failures may be correlated with one another. These factors make the system modeling and analysis complicated. As a result, making decisions in the system design or resource allocation also becomes difficult accordingly.

There is no common approach to assess computing systems. Reliability is a quantitative measure useful in this context as reliability can be broadly interpreted as the ability for a system to perform its intended function. Intensive studies on reliability models and analytical tools are carried out to improve the chance that the computing systems will perform satisfactorily in operations. As the functionality of computing operations becomes more essential, there is a greater need for a high reliability of the computing systems.

In fact, in order to increase the performance of the computing systems and to improve the development process, a thorough analysis of their reliability is needed. Based on the models and analysis, approaches to improve system reliability can be further implemented.

1.2. Computing System Reliability Concepts

In general, the basic reliability concept is defined as the probability that a system will perform its intended function during a period of running time without any failure (Musa, 1998). A failure causes the system performance to deviate from the specified performance.

A fault is an erroneous state of the system. Although the definitions of fault are different for different systems and in different situations, a fault is always an existing part in the system and it can be removed by correcting the erroneous part of the system. For the computing systems, the basic reliability concept can be adapted to some specific forms such as "software reliability", "system reliability", "service reliability", "system availability", etc., for different purposes.

Most computing systems contain software programs to achieve various computing tasks. Software reliability is an important metric to assess the software performance. Similar to the general reliability concept, software reliability is defined as the probability that the software will be functioning without failure under a given environmental condition during a specified period of time (Xie, 1991). Here, a software failure means generally the inability of performing an intended task specified by the requirement.

Software reliability is only a measurement of software program. In order to assess the computing system that may contain multiple software programs and hardware components, system reliability is commonly used. It is defined as the probability that all the tasks for which the system is desired can be successfully completed (Kumar *et al.*, 1986). Those software programs may be in parallel or serial and they may even have any arbitrarily distributed structure. The system reliability needs to be computed in a different way according to the system structure.

Some computing systems are developed to provide different services for the users. The users may only be concerned with whether the service they are using is reliable or not. From the users' point of view, service reliability is an important measure, and it is defined as the probability for a given service to be achieved successfully. This is a useful concept in service quality analysis, and it broadens the traditional reliability definition.

1.3. Approaches to Computing System Modeling

Computing system reliability is an interesting, but difficult, research area. Although there are many reliability models suggested and studied in the literature, none can be used universally, and there is no unique model which can perform well in all situations. The reason for this is that the assumptions made for each model are correct or are good approximations of the reality only in specific cases.

In the computing systems, hardware (such as computers, routers, processors, CPUs, memories, disks, etc.) provides the fundamental configurations to support computing tasks. Many traditional reliability models mainly dealt with the hardware reliability, such as Barlow & Proschan (1981), Elsayed (1996) and Blishcke & Murthy (2000).

Software is another important element in the computing systems besides the hardware. Different from the hardware, the software does not wear-out and it can be easily reproduced. Furthermore, software systems are usually debugged during testing phase so that its reliability is improving over time. Many software reliability models have been proposed for the study of software reliability, see e.g., Xie (1991), Lyu (1996), Musa (1998) and Pham (2000).

However, a computing system usually includes not only a hardware subsystem but also a software subsystem, which ought not to be separately studied. Both software and hardware failures should be integrated together in analyzing the performance of the whole system. Many reliability models for the integrated software and hardware systems have been recently presented, such as Goel & Soenjoto (1981), Siegrist (1988), Laprie & Kanoun (1992), Dugan & Lyu, (1994), Welke *et al.* (1995) and Lai *et al.* (2002). Although there are some books that contain discussion on integrated software and hardware system reliability, this book is entirely devoted to this topic and the associated issues.

Accompanying the development of network techniques, many computing systems need to communicate information through the (local or global) networks. The programs and resources of such systems are distributed all over the different sites connected by the networks. This kind of computing system is usually called distributed computing system. The performance of a distributed computing system is determined not only by the software/hardware reliability but also by the reliability of the networks for communication. Many models and algorithms have been presented for the distributed system reliability, see e.g. Hariri *et al.* (1985), Kumar *et al.* (1986), Chen & Huang (1992), Chen *et al.* (1997), Lin *et al.* (1999, 2001) and Dai *et al.* (2003a).

As a special type of the distributed computing systems, grid computing is a recently developed technique by its focus on various shared resources, large-scale networks, wide-area communications, real-time programs, diverse virtual organizations, heterogeneous platforms etc. Many experts believe that the grid computing systems and technologies will offer a second chance to fulfill the promises of the Internet, see e.g. Foster & Kesselman (1998). Although it is difficult to study due to its complexity, the reliability of the grid computing systems begins to be of concern today.

Most of reliability models for computing systems assume only two possible states of the system. In reality, many computing systems may contain more than two states (Lisnianski & Levitin, 2003), especially for those real-time systems. For example, if some computing elements in a real-time system fail, the system may still continue working but its performance should be degraded. Such a degradation state is another state between the perfect working and completely failed states. To study these types of systems, the Multi-State system reliability is also of concern recently to many researchers, e.g. Brunelle & Kapur (1999), Pourret *et al.* (1999), Levitin *et al.* (2003) and Wu & Chan (2003).

The book provides a systematic and comprehensive study of different reliability models and analytical tools for various computing systems including hardware, software, integrated software/hardware, distributed computing, grid computing, multi-state systems etc. Some interesting optimization problems for system design and resource allocation are further discussed. Many examples are used to illustrate to the use of these models.

CHAPTER 2

BASIC RELIABILITY CONCEPTS AND ANALYSIS

Reliability concepts and analytical techniques are the foundation of this book. Many books dealing with general and specific issues of reliability are available, see e.g., Barlow & Proschan (1981), Shooman (1990), Hoyland & Rausand (1994), Elsayed (1996), and Blischke & Murthy (2000). Some basic and important reliability measures are introduced in this chapter. Since computing system reliability is related to general system reliability, the focus will be on tools and techniques for system reliability modeling and analysis. Since Markov models will be extensively used in this book, this chapter also introduces the fundamentals of Markov modeling. Moreover, Nonhomogeneous Poisson Process (NHPP) is widely used in reliability analysis, especially for repairable systems. Its general theory is also introduced for the reference.

2.1. Reliability Measures

Reliability is the analysis of failures, their causes and consequences. It is the most important characteristic of product quality as things have to be working satisfactorily before considering other quality attributes. Usually, specific performance measures can be embedded into reliability analysis by the fact that if the performance is below a certain level, a failure can be said to have occurred.

2.1.1. Definition of reliability

The commonly used definition of reliability is the following.

Definition 2.1. *Reliability* is the probability that the system will perform its intended function under specified working condition for a specified period of time.

Mathematically, the reliability function $R(t)$ is the probability that a system will be successfully operating without failure in the interval from time 0 to time t,

$$R(t) = P(T > t), \quad t \geq 0 \tag{2.1}$$

where T is a random variable representing the failure time or time-to-failure.

The failure probability, or unreliability, is then

$$F(t) = 1 - R(t) = P(T \leq t)$$

which is known as the distribution function of T.

If the time-to-failure random variable T has a density function $f(t)$, then

$$R(t) = \int_t^\infty f(x)dx$$

The density function can be mathematically described as $\lim_{\Delta t \to 0} P(t < T \leq t + \Delta t)$. This can be interpreted as the probability that the failure time T will occur between time t and the next interval of operation, $t + \Delta t$. The three functions, $R(t)$, $F(t)$ and $f(t)$ are closely related to one another. If any of them is known, all the others can be determined.

2.1.2. Mean time to failure (MTTF)

Usually we are interested in the expected time to next failure, and this is termed mean time to failure.

Definition 2.2. The *mean time to failure (MTTF)* is defined as the expected value of the lifetime before a failure occurs.

Suppose that the reliability function for a system is given by $R(t)$, the MTTF can be computed as

$$\text{MTTF}=\int_0^\infty t\cdot f(t)dt=\int_0^\infty R(t)dt \qquad (2.2)$$

Example 2.1. If the lifetime distribution function follows an exponential distribution with parameter λ, that is, $F(t)=1-\exp(-\lambda t)$, the MTTF is

$$\text{MTTF}=\int_0^\infty R(t)dt=\int_0^\infty \exp(-\lambda t)dt=\frac{1}{\lambda} \qquad (2.3)$$

This is an important result as for exponential distribution. MTTF is related to a single model parameter in this case. Hence, if MTTF is known, the distribution is specified.

■

2.1.3. Failure rate function

The failure rate function, or hazard function, is very important in reliability analysis because it specifies the rate of the system aging. The definition of failure rate function is given here.

Definition 2.3. The *failure rate function* $\lambda(t)$ is defined as

$$\lambda(t) = \lim_{\Delta t \to 0} \frac{R(t) - R(t + \Delta t)}{\Delta t R(t)} = \frac{f(t)}{R(t)} \tag{2.4}$$

The quantity $\lambda(t)dt$ represents the probability that a device of age t will fail in the small interval from time t to $t + dt$. The importance of the failure rate function is that it indicates the changing rate in the aging behavior over the life of a population of components. For example, two designs may provide the same reliability at a specific point in time, but the failure rate curves can be very different.

Example 2.2. If the failure distribution function follows an exponential distribution with parameter λ, then the failure rate function is

$$\lambda(t) = \frac{f(t)}{R(t)} = \frac{\lambda \cdot \exp(-\lambda t)}{\exp(-\lambda t)} = \lambda \tag{2.5}$$

This means that the failure rate function of the exponential distribution is a constant. In this case, the system does not have any aging property. This assumption is usually valid for software systems. However, for hardware systems, the failure rate could have other shapes.

■

2.1.4. Maintainability and availability

When a system fails to perform satisfactorily, repair is normally carried out to locate and correct the fault. The system is restored to operational effectiveness by making an adjustment or by replacing a component.

Definition 2.4. *Maintainability* is defined as the probability that a failed system will be restored to a functioning state within a given period of time when maintenance is performed according to prescribed procedures and resources.

Generally, maintainability is the probability of isolating and repairing a fault in a system within a given time. Maintenance personnel have to work with system designers to ensure that the system product can be maintained cost effectively.

Let T denote the time to repair or the total downtime. If the repair time T has a density function $g(t)$, then the maintainability, $V(t)$, is defined as the probability that the failed system will be back in service by time t, i.e.,

$$V(t) = P(T \leq t) = \int_0^t g(x)dx$$

An important measure often used in maintenance studies is the mean time to repair (MTTR) or the mean downtime. MTTR is the expected value of the repair time.

Another important reliability related concept is system availability. This is a measure that takes both reliability and maintainability into account.

Definition 2.5. The *availability* function of a system, denoted by $A(t)$, is defined as the probability that the system is available at time t.

Different from the reliability that focuses on a period of time when the system is free of failures, availability concerns a time point at which the system does not stay at the failed state. Mathematically,

$$A(t) = \Pr(\text{System is up or available at time instant } t)$$

The availability function, which is a complex function of time, has a simple steady-state or asymptotic expression. In fact, usually we are mainly concerned with systems running for a long time. The steady-state or asymptotic availability is given by

$$A=\lim_{t\to\infty} A(t)=\frac{\text{System up time}}{\text{System up time}+\text{System down time}}=\frac{\text{MTTF}}{\text{MTTF}+\text{MTTR}}$$

The mean time between failures (MTBF) is another important measure in repairable systems. This implies that the system has failed and has been repaired. Like MTTF and MTTR, MTBF is an expected value of the random variable time between failures. Mathematically, MTBF=MTTR+ MTTF.

Example 2.3. If a system has a lifetime distribution function $F(t)=1-\exp(-\lambda t)$ and a maintainability function $V(t)=1-\exp(-\mu t)$, then MTTF= $1/\lambda$ and MTTR=$1/\mu$. The MTBF is the sum of MTTF and MTTR and the steady-state availability is

$$A=\frac{\text{MTTF}}{\text{MTTR}+\text{MTTF}}=\frac{1/\lambda}{1/\lambda+1/\mu}=\frac{\mu}{\lambda+\mu}$$

■

2.2. Common Techniques in Reliability Analysis

There are many techniques in reliability analysis. The most widely used techniques in computing systems are reliability block diagrams, network diagrams, fault tree analysis and Monte Carlo simulation, which will be introduced in the following sections. Another popular and important analytical tool, Markov model, will be introduced in Section 2.3 since it is the main technique used in this book.

2.2.1. Reliability block diagram

A reliability block diagram is one of the conventional and most common tools of system reliability analysis. A major advantage of using the reliability block diagram approach is the ease of reliability expression and evaluation.

A reliability block diagram shows the system reliability structure. It is made up of individual blocks and each block corresponds to a system module or function. Those blocks are connected with each other through certain basic relationships, such as series and parallels. The series relationship between two blocks is depicted by Fig. 2.1 (a) and parallel by Fig. 2.1 (b).

(a) Series blocks (b) Parallel blocks

Fig. 2.1. Basic relationships between two blocks.

Suppose that the reliability of a block for module i is known or estimated, and it is denoted by R_i. Assuming that the blocks are independent from a reliability point of view, the reliability of a system with two serially connected blocks is

$$R_s = R_1 \cdot R_2 \tag{2.6}$$

and that of a system with two parallel blocks is

$$R_p = 1 - \prod_{i=1}^{2}(1 - R_i) \tag{2.7}$$

The blocks in either series or parallel structure can be merged into a new block with the reliability expression of the above equations. Using such combinations, any parallel-series system can be eventually merged to one block and its reliability can be easily computed by repeatedly using those equations.

Example 2.4. A parallel-series system consists of five modules whose reliability block diagram is shown as Fig. 2.2(a). The parallel blocks can be merged as shown by Fig. 2.2(b). It can be further merged into one block simply through the series expression (2.6). The combined reliability expression is given under the new blocks.

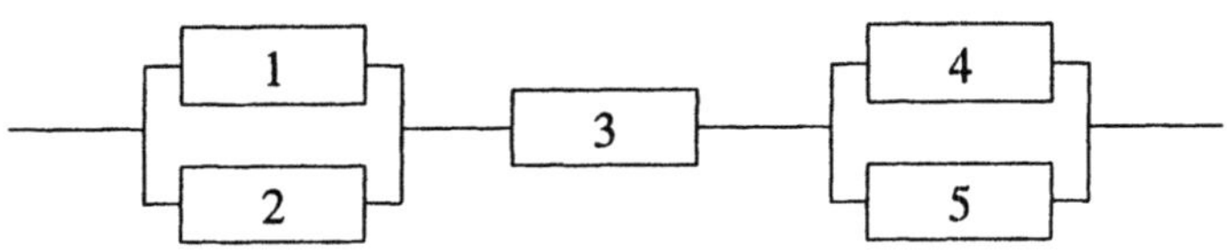

(a) Five parallel-series connected modules

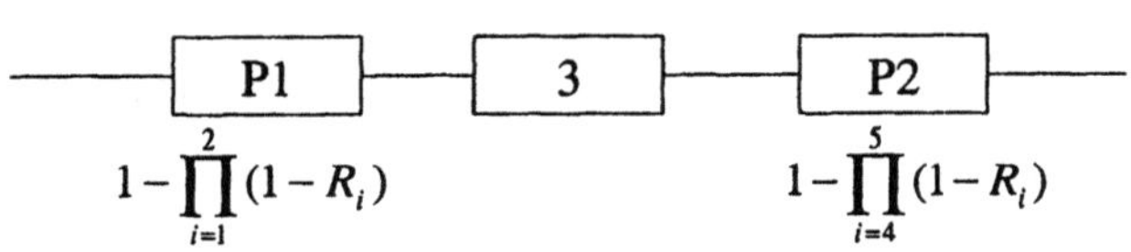

(b) The merged blocks

Fig. 2.2. Reliability computation of a parallel-series system.

■

Furthermore, a library for reliability block diagrams can be constructed in order to include other configurations or relationships. Additional notational description is needed and specific formulas for evaluating these blocks must be obtained and added to the library. One such example is the simple k-out-of-n in the following.

Example 2.5. A *k*-out-of-*n* system requires that at least *k* modules out of a total of *n* must be operational in order for the system to be working. Usually a voter is needed, see Fig. 2.3.

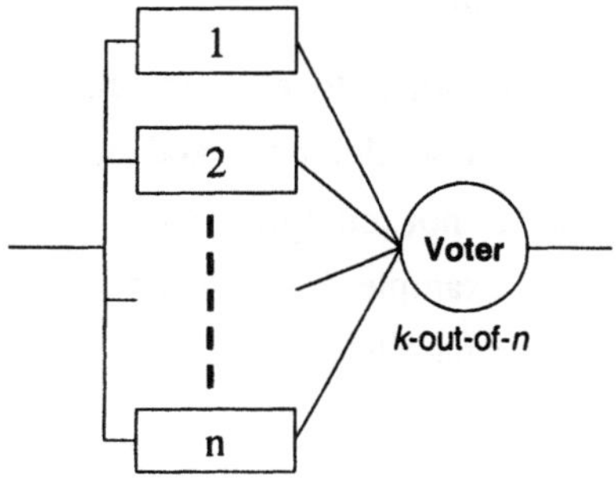

Fig. 2.3. The *k*-out-of-*n* configuration representing by blocks.

If the voter is perfect and all the modules have reliability R, the formula to evaluate the reliability of these blocks, which can be obtained via conditioning or binomial distribution (Barlow & Proschan, 1981), is

$$R_v = \sum_{i=k}^{n} \frac{n!}{i!(n-i)!} R^i (1-R)^{n-i} \tag{2.8}$$

■

A majority voting system requires more than half of modules to be operational. The reliability of such a system is given by

$$R_v = \sum_{i=[n/2]+1}^{n} \frac{n!}{i!(n-i)!} R^i (1-R)^{n-i}$$

where $[X]$ denotes the largest integer that is less than or equal to *X*.

2.2.2 Network diagram

Network diagrams are commonly used in representing communication networks consisting of individual links. Most network applications are in the communication domain. The computation of network reliability is the primary application of network diagrams.

The purpose of a network is to execute programs by connecting different sites that contain processing elements and resources. For simple network diagrams, computation is not complex and reliability block diagrams can alternatively be used. For example, Fig. 2.4 shows the network diagrams that are connected through series or parallel links.

(a) Series connected links (b) Parallel connected links

Fig. 2.4. Network diagram representing series and parallel two links.

Fig. 2.4 can alternatively be represented by the reliability block diagrams if we view each link as a block, depicted by Fig. 2.1.

The choice of reliability block diagram or network diagram depends on the convenience of their usage and description for certain specific problems. Usually, the reliability block diagram is mainly used in a modular system that consists of many independent modules and each module can be easily represented by a reliability block. The network diagram is often used in networked system where processing nodes are connected and communicated through links, such as the distributed computing system, local/wide area networks and the wireless communication channels, etc.

2.2.3. Fault tree analysis

Fault tree analysis is a common tool in system safety analysis. It has been adapted in a range of reliability applications.

A fault tree diagram is the underlying graphical model in fault tree analysis. Whereas the reliability block diagram is mission success oriented, the fault tree shows which combinations of the component failures will result in a system failure. The fault tree diagram represents the logical relationships of 'AND' and 'OR' among diverse failure events. Various shapes represent different meanings. In general, four basic shapes corresponding to four relationships are depicted by Fig. 2.5.

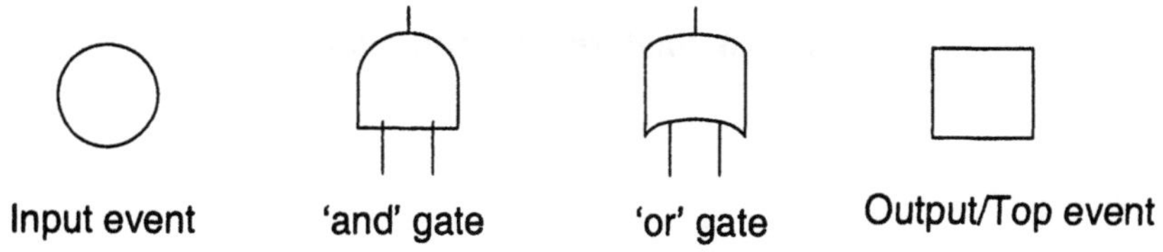

Fig. 2.5. Basic shapes of fault tree diagram.

Since any logical relationships can be transformed into the combinations of 'AND' and 'OR' relationships, the status of output/top event can be derived by the status of input events and the connections of the logical gates.

Example 2.6. An example of a fault tree diagram corresponding to the reliability block diagram in Example 2.4 is shown by Fig. 2.6. As the fault tree shows, the *top-event* of the system fails if both module 1 and 2 fail, or module 3 fails, or both module 4 and 5 fail.

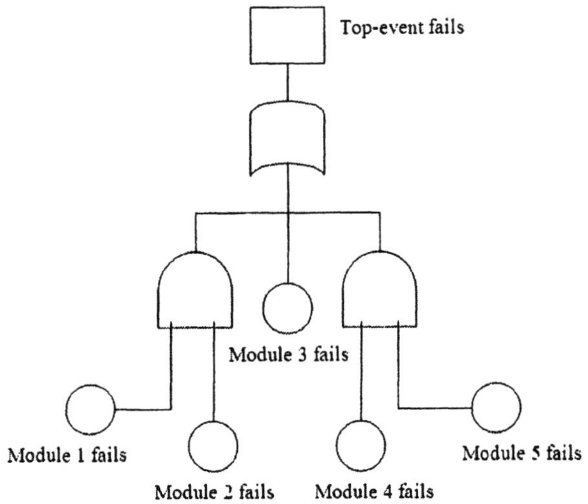

Fig. 2.6. Fault tree for five modules.

■

A fault tree diagram can describe the fault propagation in a system. However, complex systems may exhibit much more complex failure behavior, including multiple failure modes and dependent failure modes. These failures will have different effects on the mission outcome. The basic fault tree analysis does not support this type of modeling. Moreover, repair and maintenance are two important operations in system analysis that cannot be expressed easily using a fault tree formulation.

2.2.4. Monte Carlo simulation

In a Monte Carlo simulation, a reliability model is evaluated repeatedly using parameter values drawn from a specific distribution. The Monte Carlo simulation is often used to evaluate the MTBF for complex systems. Here, the following steps apply:

1) Simulate random numbers for each random variable needed in the simulation model.

2) Evaluate the desired function.

3) Repeat steps 1 and 2 a total of n times, to obtain n samples of the desired function. For example, the system failure times will be $T(1)$, $T(2),\ldots,T(n)$.

4) Estimate the desired parameter. For example, the expected value of the system failure time can be obtained from

$$E(T) = \overline{MTBF} = \frac{1}{n}\sum_{i=1}^{n} T(i)$$

5) Obtain an estimate of the precision of the estimate, such as the sample standard deviation of the estimated value.

Monte Carlo simulation can handle a variety of complex system configurations and failure rate models. However, Monte Carlo simulation usually requires the development of a customized program, unless the system configuration fits a standard model. It also requires lengthy computer runs if accurate and converging computations are desired.

2.3. Markov Process Fundamentals

Markov model is another widely used technique in reliability analysis. It overcomes most disadvantages of other techniques and is more flexible to be implemented in reliability analysis for various computing systems, which will be applied in the later chapters.

2.3.1. Stochastic processes

When we examine the evolution of a process governed by the rules of probability, we observe a *stochastic process*. The study of stochastic processes involves the analysis of a collection of random variables, their interdependence, their change over time, and limiting behavior, among others (Ross, 2000).

In the study of stochastic processes, it is useful to establish two distinct categories:

1) *Stationary*: A stationary process is one for which the distribution remains the same over time.
2) *Evolutionary* (*Nonstationary*): An evolutionary process can be defined as one that is not stationary and the process evolves with time.

Almost all systems are dynamic in nature. Markov model is a powerful tool to solve such dynamic problems. Its stochastic process is a sequence of outcomes X_t, where t takes value from a *parameter space T*.

If the parameter space T is discrete and countably finite, the sequence is called a discrete-time process and is denoted by $\{X_n\}$ where n=1,2,.... The index n identifies the steps of the process. On the other hand, if the parameter space T is continuous or uncountable, the sequence is called a continuous-time process and is denoted by $\{X_t, t \in [0,\infty)\}$.

The set of all possible and distinct outcomes of all experiments in a stochastic process is called its *state space* and normally is denoted by Ω. Its elements are called the *states*. If the state space Ω is discrete, then the process is called a discrete-state process. Otherwise, it is called continuous-state process.

2.3.2. Standard Markov models

There are four types of standard Markov models corresponding to four types of Markov processes classified according to their state-space and time characteristics as Table 2.1 shows below.

Table 2.1. Four types of Markov processes.

Type	*State Space*	*Time Space*
1	Discrete	Discrete
2	Discrete	Continuous
3	Continuous	Discrete
4	Continuous	Continuous

The standard Markov models satisfy the *Markov property*, which is defined here.

Definition 2.6. For a stochastic process that possesses *Markov property*, the probability of any particular future behavior of the process, when its current state is known exactly, is not changed by additional information concerning its past behavior.

These four Markov models are described in more details in the following sections.

Discrete-Time Markov chain

The discrete-state process is referred to as *chain*, so the discrete-state and discrete-time Markov process is usually called discrete time Markov chain (DTMC).

A general discrete-time chain is a sequence of discrete random variables $\{X_n, n=1,2,...\}$, in which X_{k+1} is dependent on all previous outcomes $X_0, X_1,..., X_k$. The analysis of this type of chain can easily become unmanageable, especially for long-term evaluation. Fortunately, in many practical situations, the influence of the earlier outcomes on its future one tends to diminish rapidly with time.

For mathematical tractability, we can assume that X_{n+1} is dependent only on *i* previous outcomes, where $i \geq 1$ is a fixed and finite number. In this case, deriving $\Pr\{X_{n+1}=j\}$ requires only the information about the previous *i* outcomes (from step *n-i*+1 to step *n*), i.e.,

$$\Pr\{X_{n+1}=j \mid X_0=i_0, X_1=i_1,..., X_n=i\}$$

$$=\Pr\{X_{n+1}=j \mid X_{n-i+1}=i_{n-i+1}, X_{n-i+2}=i_{n-i+2},..., X_n=i\} \tag{2.9}$$

We call this type of chain a *Markov chain of order i*.

We usually refer to the first-order Markov chain simply as a Markov chain. For these chains, only their present (at time *n*) has any influence on their future (at time *n*+1). In other words, for all *n*>0,

$$\Pr\{X_{n+1}=j \mid X_0=i_0, X_1=i_1,..., X_n=i\}=\Pr\{X_{n+1}=j \mid X_n=i\} \tag{2.10}$$

The essential characteristic of such a Markov process can be thought of as *memoryless*.

For the right-hand side of the above equation, it is assumed that the state space Ω under consideration is either finite or countably infinite. Define

$$p_{ij}(n,n+1)=\Pr\{X_{n+1}=j \mid X_n=i\},\ n=0,1,... \tag{2.11}$$

The conditional probability $p_{ij}(n,n+1)$ is called the (one-step) *transition probability* from state *i* to state *j* at time *n*. The *m*-step transition probabilities at time *n* are defined by

$$p_{ij}(n, n+m) = \Pr\{X_{n+m} = j \mid X_n = i\}\text{, } n\text{=0,1,...} \tag{2.12}$$

and the corresponding m-step transition matrix at time n is $\mathbf{P}(n,n+m)$. The transition matrix should satisfy,

$$\mathbf{P}(m,n) = \mathbf{P}(m,l)\, \mathbf{P}(l,n), \quad m \le l \le n \tag{2.13}$$

or, equivalently,

$$p_{ij}(m,n) = \sum_k p_{ik}(m,l)\, p_{kj}(l,n), \quad m \le l \le n \tag{2.14}$$

This equation is known as the *Chapman-Kolmogorov* equation (Ross, 2000).

Example 2.7. Suppose that a computing system has three states after each run. The states are perfect, degraded, and failed states denoted by state 1, 2 and 3. The state of the current run will just affect the state of the next run. The matrix of one step transition probability is

$$P = \begin{bmatrix} 0.7 & 0.2 & 0.1 \\ 0.3 & 0.5 & 0.2 \\ 0.1 & 0.3 & 0.6 \end{bmatrix}$$

This is a discrete time, discrete state Markov chain (DTMC) that is depicted by the transition graph in Fig. 2.7.

According to the Chapman-Kolmogorov equation, the two-step transition matrix can be obtained as

$$P(0,2) = P(0,1) \times P(1,2) = \begin{bmatrix} 0.7 & 0.2 & 0.1 \\ 0.3 & 0.5 & 0.2 \\ 0.1 & 0.3 & 0.6 \end{bmatrix}^2 = \begin{bmatrix} 0.56 & 0.27 & 0.17 \\ 0.38 & 0.37 & 0.25 \\ 0.22 & 0.35 & 0.43 \end{bmatrix}$$

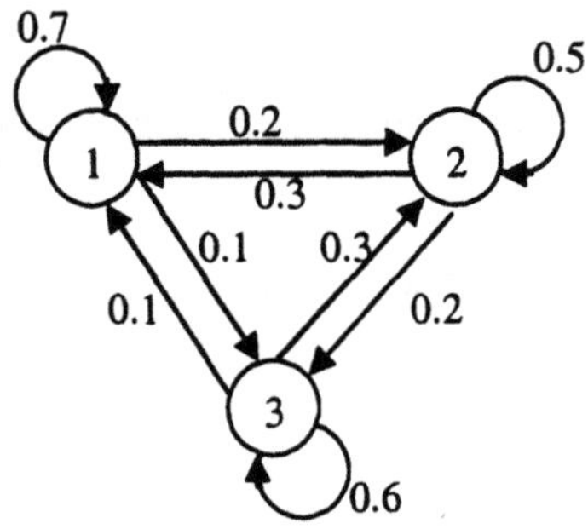

Fig. 2.7. DTMC for the three-state system transitions.

Thereafter, if the system initially stays at a perfect state, then the probability that the system still stays at that state after 2 runs should be $p_{11}(0,2)=0.56$. The four-step transition matrix is

$$P(0,4)=P(0,2)\times P(2,4)=\begin{bmatrix}0.56 & 0.27 & 0.17\\ 0.38 & 0.37 & 0.25\\ 0.22 & 0.35 & 0.43\end{bmatrix}^2=\begin{bmatrix}\dots & \dots & 0.236\\ \dots & \dots & \dots\\ \dots & \dots & \dots\end{bmatrix}$$

The probability that the system does not stay at the failed state after 4 runs is

$$1-p_{13}(0,4)=1-0.236=0.764$$

■

Continuous-time Markov chain

Similar to the case of DTMC, the discrete-state and continuous-time Markov process is usually called the continuous time Markov chain. Let the time space $T=[0,\infty)$ be an index set and consider a continuous-time stochastic process $\{X(t),t\geq 0\}$ taking values on the discrete state space Ω. We say that the process $\{X(t), t \geq 0\}$ is a Markov chain in continuous time if, for each $s \geq 0$, $t > 0$ and each set *A*, we have

$$\Pr\{X(t+s) \in A \mid X(u), 0 \leq u \leq s\} = \Pr\{X(t+s) \in A \mid X(s)\}$$

Specifically, if, for each $s \geq 0$, $t > 0$ and each $i, j \in \Omega$, and every history $x(u)$, $0 \leq u \leq s$,

$$\Pr\{X(t+s) = j \mid X(s) = i, X(u), 0 \leq u \leq s\} = \Pr\{X(t+s) = j \mid X(s) = i\} \quad (2.15)$$

then the process $\{X(t)\}$ is called a *continuous-time Markov chain* (CTMC).

A CTMC is a stochastic process having the Markov property that the conditional distribution of the future state, given the present state and all past states, depends only on the present state and is independent of the past. Also, define

$$p_{ij}(s,t) = \Pr\{X(t) = j \mid X(s) = i\}, \quad 0 \leq s < t \quad (2.16)$$

The conditional probability $p_{ij}(s,t)$ is called the *transition probability function* from state *i* to state *j* and the matrix **P**(*s*,*t*) is called the *transition matrix function.*

Similar to the DTMC, we have the Chapman-Kolmogorov equation as

$$p_{ij}(s,t) = \sum_k p_{ik}(s,u) p_{kj}(u,t), \quad 0 \leq s < u < t \quad (2.17)$$

In matrix notation, this can be written as

$$\mathbf{P}(s,t) = \mathbf{P}(s,u)\, \mathbf{P}(u,t), \quad 0 \leq s < u < t \quad (2.18)$$

The above equation can be compared with its discrete-time counter-part (2.13) or (2.14).

When the transition probability functions $p_{ij}(s,t)$ depend only on the difference $\Delta t = t - s$, i.e.,

$$p_{ij}(\Delta t) = \Pr\{X(\Delta t + s) = j \mid X(s) = i\}, \quad 0 \leq s < t \text{, for all } i, j \in \Omega,$$

the continuous-time Markov chain $\{X(t)\}$ is said to be *homogeneous*. For any homogeneous Markov chain, the Chapman-Kolmogorov equation is expressed as

$$p_{ij}(s+t)=\sum_{k} p_{ik}(s)p_{kj}(t), \quad s,t>0 \tag{2.19}$$

This can be written in matrix form as

$$\mathbf{P}(s+t)=\mathbf{P}(s)\,\mathbf{P}(t), \qquad s,t>0 \tag{2.20}$$

where $\mathbf{P}(t)=\{p_{ij}(t)\}$ which satisfies

$$\mathbf{P}(t\text{-}s)=\mathbf{P}(s,t) \text{ for } t>s\geq 0$$

As given by Kijima (1997, p.174), the derivative of $\mathbf{P}(t)$ is defined as

$$\mathbf{P}'(t)=[\mathrm{P}(t+h)-\mathrm{P}(t)]\cdot\left[\int_0^h \mathrm{P}(u)du\right]^{-1} \tag{2.21}$$

which shows that $\mathbf{P}(t)$ is infinitely differentiable with respect to $t>0$.

Define $\mathbf{Q}\equiv\ \mathbf{P}'(0+)$. The matrix $\mathbf{Q}=\{q_{ij}\}$ is called *infinitesimal generator*, or *generator* for short. This is of fundamental importance in the theory of CTMC. Since $\mathbf{P}(0)=\mathbf{I}$, we have

$$q_{ij}=\begin{cases} \lim\limits_{h\to 0+}\dfrac{p_{ij}(h)}{h}\geq 0, & i\neq j \\ \\ \lim\limits_{h\to 0+}\dfrac{p_{ii}(h)-1}{h}\leq 0, & i=j \end{cases} \tag{2.22}$$

Since $\mathbf{P}(t)$ is differentiable, it follows from (2.22) that

$$\mathbf{P}'(t)=\mathbf{P}(t)\,\mathbf{Q}, t\geq 0 \tag{2.23}$$

which are the systems of ordinary linear differential equations. The former is known as the backward Kolmogorov equation and the latter as the forward Kolmogorov equation (Ross, 2000).

Example 2.8. Suppose that a computing system has two states: Good and Failed, denoted by 1 and 2, respectively. Suppose that the transition from state i to j follow a continuous time distribution, say the exponential distribution,

$$F_{ij}(t) = 1 - \exp(-\lambda_{ij} t),\ i,j=1,2$$

The CTMC is depicted in Fig. 2.8.

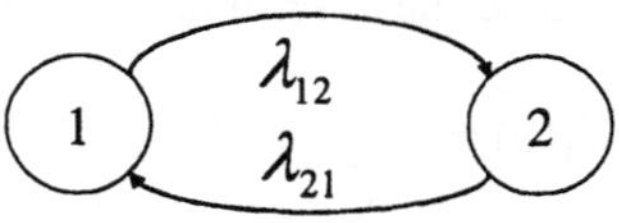

Fig. 2.8. CTMC for the two-state system.

From the exponential distribution, we have

$$p_{ij}(h) = 1 - \exp(-\lambda_{ij} h),\quad i \neq j \tag{2.24}$$

Then, q_{ij} can be written as Eq. (2.22) for $(i \neq j)$

$$\begin{aligned} q_{ij} &= \lim_{h\to 0+} \frac{p_{ij}(h)}{h} = \lim_{h\to 0+} \frac{1-\exp(-\lambda_{ij} h)}{h} \\ &= \lim_{h\to 0+} \frac{\exp(-\lambda_{ij} t) - \exp\{-\lambda_{ij}(t+h)\}}{h \cdot \exp(-\lambda_{ij} t)} \end{aligned} \tag{2.25}$$

Let $R_{ij}(t) = \exp(-\lambda_{ij} t)$. We have,

$$q_{ij} = \lim_{h \to 0+} \frac{R_{ij}(t) - R_{ij}(t+h)}{h \cdot R_{ij}(t)} = \frac{1}{R_{ij}(t)}\left[-\frac{d}{dt}R_{ij}(t)\right] = \lambda_{ij} \tag{2.26}$$

This solution is useful, and it implies that for exponential distribution, the q_{ij} is equal to its rate.

Then, the Chapman-Kolmogorov equation for Fig. 2.8 can be written as

$$P_1'(t) = \lambda_{21}P_2(t) - \lambda_{12}P_1(t) \tag{2.27}$$

and

$$P_2'(t) = \lambda_{12}P_1(t) - \lambda_{21}P_2(t)\,. \tag{2.28}$$

With the initial condition (assume the system initially stays at the good state)

$$P_1(0) = 1,\ P_2(0) = 0 \tag{2.30}$$

we obtain the availability function as

$$A(t) = P_1(t) = \frac{\lambda_{12}}{\lambda_{21} + \lambda_{12}}\exp\{-(\lambda_{21} + \lambda_{12})t\} + \frac{\lambda_{21}}{\lambda_{21} + \lambda_{12}} \tag{2.31}$$

■

Discrete Time, Continuous State

The discrete-time continuous-state Markov model is applicable if there are discrete changes in time in an environment where the states of the system are continuous over a specified range.

It is easy to see how the concept could be applied to the component parameter drift problem. However, little work has been done in this area, and multi-parameter modeling and computation remain a difficult problem. There are two possible reasons: numerical data are seldom available, and the solution of the resulting partial differential equations is more complex.

Continuous Time, Continuous State

The conventional diffusion equations fall in this category of continuous-time and continuous-state Markov models. Usually when we talk about the system state space, we attempt to describe it in fixed terms. In reliability, we talk about fully operational systems or failed systems. Once we introduce the concept of degraded operability, it is easy to imagine a continuum of physical states in which the system can exist. There could be some other advanced applications. However, the evaluation of these equations will be costly and more involved.

Since little work has been done in the area of the continuous state (Type 3 and 4 in Table 2.1), the continuous-state Markov process will not be discussed in this book. For details about them, the readers can refer to Kijima (1997).

2.3.3. Some non-standard Markovian models

Some important aspects of system behavior cannot be easily captured in certain types of the above Markov models. The common characteristic these problems share is that the Markov property is not valid at all time instants. This category of problems is jointly referred to as *non-Markovian* models and can be analyzed using several approaches, see e.g., Limnios & Oprisan (2000).

Markov renewal sequence

We first introduce the *renewal process*. Let $S_0 < S_1 < S_2 < \dots$ be the time instants of successive events to occur. The sequence of non-negative independent and identically distributed random variables, **S**= $\{S_n - S_{n-1}; n = 1,2,\dots\}$ is a *renewal process*.

The idea of having the times $S_n - S_{n-1}$ depend on a state which can be generalized. We can assume that there is a set of states Ω, which can be thought of as the set 0, 1,..., as before. The state at S_n is given by $X_n \in \Omega$. The chain X_n now forms a process on its own. In particular, they may form a DTMC. The points S_n ; n=0,1,2,..., are called *Markov regeneration epochs*, or *Markov renewal moments*. Together with the states of the *embedded Markov chain* X_n, they define a Markov renewal sequence.

Definition 2.7. The bivariate stochastic process (*X*,*S*) = $\{X_n, S_n; n = 1,2,...\}$ is a *Markov renewal sequence* provided that

$$\Pr\{X_{n+1} = j, S_{n+1} - S_n \le t | X_0,...,X_n; S_0,...,S_n\}$$

$$= \Pr\{X_{n+1} = j, S_{n+1} - S_n \le t | X_n\}, \quad n = 0,1,2,..., j \in \Omega, t \ge 0 \tag{2.32}$$

The random variables S_n are the regeneration epochs, and the X_n are the states at these epochs.

Markov renewal sequences are embedded into Markov Renewal Models. Markov renewal models can be classified into two categorizations called semi-Markov model and Markov regenerative model.

Semi-Markov process

A possible generalization of the CTMC is to allow the *holding time* to follow general distributions. That is, by letting $F_i(t)$ be the holding-time distribution when the process is in state *i*, we can construct a stochastic process $\{X(t)\}$ as follows. If $X(0) = i$, then the process stays in state *i* for a time with distribution function $F_i(t)$. At the end of the holding time, the process moves to state *j*, which can be equal to *i*, according to the Markovian law **P**=$\{p_{ij}\}$. The process stays in state j for a time with distribution function $F_j(t)$ and then moves to some state according to **P**. Under some regularity conditions, we can construct a stochastic process by repeating the above procedure.

We can introduce more dependent structure into the holding times. Namely, when $X(0)=i$, we choose the next state j and the holding time simultaneously according to a joint distribution $F_{ij}(t)$. Given the next state j, the holding-time distribution is given by $F_{ij}(t)/F_{ij}(\infty)$. After the holding time, a transition to state j occurs. At the same time, the next state k as well as the holding time is determined according to a joint distribution $F_{jk}(t)$. A stochastic process constructed in this way is called a *semi-Markov process.*

Definition 2.8. Let Ω denote the state space and let $\{Y_n\}$ be a sequence of random variables taking values on Ω. Let $\{V_n\}$ be a sequence of random variables taking values on $R_+ = [0,\infty)$ and let

$$\tau_n = \sum_{k=0}^{n-1} V_k \text{ , } n=1,2,\ldots, \text{ with } \tau_0 \equiv 0$$

We define $\tau(t) = \max\{n : \tau_n \le t\}$, $t \ge 0$, the renewal process associated with $\{V_n\}$. Thereafter, with the above notation, suppose that

$$\Pr\{Y_{n+1}=j, V_n \le t \mid Y_0,\cdots Y_n = i; V_0,\cdots V_{n-1}\} = \Pr\{Y_{n+1}=j, V_n \le t \mid Y_n = i\} \quad (2.33)$$

for all n=0,1,...; $i, j \in \Omega$, and $t \ge 0$. Then the stochastic process $\{X(t)\}$ defined by $X(t) = Y_{\tau(t)}$, $t \ge 0$, is called a *semi-Markov process.*

For a semi-Markov process, the time distribution $F_{ij}(t)$ satisfies the following equation.

$$F_{ij}(t) = \sum_k p_{ik} p_{kj} \cdot F_{ik}(t) \otimes F_{kj}(t) \quad (2.34)$$

where '$\otimes$' denotes the convolution of the two functions, defined as

$$F(t)\otimes G(t)=\int_0^t F(s)G(t-s)ds. \tag{2.35}$$

Using the Laplace-Stieltjes Transform, the above equation can be simplified as

$$\tilde{F}_{ij}(s)=\sum_k p_{ik}p_{kj}\tilde{F}_{ik}(s)\tilde{F}_{kj}(s)$$

where $\tilde{F}_{ij}(s)$ is the Laplace-Stieltjes transform of $F_{ij}(t)$.

Markov regenerative process

The Markov regenerative model combines the Markov regenerative process into its modeling. A stochastic process $Z=\{Z_t;t\geq 0\}$ with the state space Ω is called regenerative if there exist time points at which the process probably restarts itself. The formal definition of Markov regenerative process is given now.

Definition 2.9. A *Markov regenerative process* is defined as a stochastic process $\{Z_t;t\geq 0\}$, $Z_t\in\Omega$ with an embedded Markov regenerative process (*X*, *S*), $X_n\in$ F, which has the additional property that all conditional finite distributions of $\{Z_{S_n+t};t\geq 0\}$, given $\{Z_u;0\leq u\leq S_n,X_n=i\}$, are the same as those of $\{Z_t;t\geq 0\}$, given $X_0=i$.

As a special case, the definition implies that for $i\in F$, $j\in\Omega$,

$$\Pr\{Z_{S_n+t}=j|Z_u,0\leq u\leq S_n,X_n=i\}=\Pr\{Z_t=j|X_0=i\} \tag{2.36}$$

The expression in Eq. (2.36) implies that the Markov regenerative process does not have the Markov property in general, but there is a sequence of embedded time points $S_0, S_1, ..., S_n, ...$, such that the states $X_0, X_1, ..., X_n, ...$ realized at these points satisfy the Markov property. It also implies that the future of the process Z from time $t = S_n$ onwards depends on the past $\{Z_u, 0 \le u \le S_n\}$ only through X_n.

Different from the semi-Markov processes, state changes in Markov regenerative process may occur between two consecutive Markov regeneration epochs. An example of Markov regenerative process is illustrated below.

Example 2.9. Suppose that a system has two states: 0 and 1 (good and failed). When the system fails, it is restarted immediately. After restarting, the system may stay at the good state (with the probability p) or failed state again (with the probability $1-p$). When the system stays at a good state, it may fail with a failure rate λ. Then the process is a Markov regenerative process, where the restarting points are regeneration epochs.

Given the initial state is the first restart state, the Markov regenerative process is depicted by Fig. 2.9 in which state S_i is the i:th restart point and G_i is the good state between S_i and S_{i+1}.

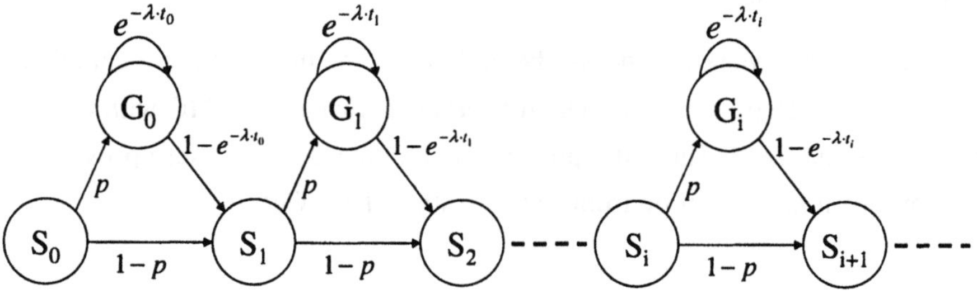

Fig. 2.9. Markov regenerative processes of Example 2.9.

■

2.3.4. General procedure of Markov modeling

A Markov process is characterized by its state space together with the transition probabilities over time between these states. The basic steps in the modeling and analysis are described in the following.

Setting up the model

In the first step, a Markov state diagram can be developed by determining the system states and the transitions between these states. It also includes labeling the states such as operational, degraded, or failed. There could be several states in the degraded category.

The state diagrams depict all possible internal relationships among states and define the allowable transitions from one state to another. In general, the state diagram is made up of nodes and links, where the nodes represent the different states and the links represent the transition between the connected two states.

For DTMC, the time between the two states is discrete, which is usually set as 1 unit. On the other hand, the time between the two states is continuous for CTMC. The Markov chain can be constructed by drawing a state diagram that is made up of the units.

For the semi-Markov process, the building of the model is more complex for it contains two steps. First, the state diagram is drawn as a DTMC with transition probability matrix **P**. Then, the process in continuous time is set up by making the time spent in a transition from state i to state j have Cdf $F_{i,j}(t)$.

Chapman-Kolmogorov equations

The second step converts the Markov state diagram developed in the preceding step into a set of equations. The well known equation for Markov models is the Chapman-Kolmogorov equations, see e.g. Trivedi (1982).

Solving the equations

Solving the state equations is sometimes complicated. An analytical solution of the state equations is feasible only for simple problems. Fortunately, a number of solution techniques exist, such as analytical solution, Laplace-Stieltjes transforms, numerical integration and computer-assisted evaluation, which can simplify this task, see e.g. Pukite & Pukite (1998, pp. 119-136).

The use of Laplace-Stieltjes transforms in engineering is well known, see Gnedenko & Ushakov (1995) for details. Important applications are in control system stability evaluation, circuit analysis, and so on. Laplace-Stieltjes transforms provide a convenient way of solving simpler models. Solution of the Markov state equations using this approach involves two steps:

a) State equations are transformed to their Laplace counterparts.

b) The resulting equations are inverted to obtain their time-domain solutions.

If the mission times are short and if the transition rates are small, then approximations can be used that may meet the accuracy requirements. An example is as follows.

Example. 2.10. Consider that a state diagram can be expressed as a sequence of transitions, as shown in Fig. 2.10.

Fig. 2.10. ***N*-state Markov diagram.**

The state probability for the last state can be given in Laplace-Stieltjes transform by

$$P_N(s) = \frac{\lambda_1 \lambda_2 \cdots \lambda_N}{(s+\lambda_1)(s+\lambda_2)\cdots(s+\lambda_N)s}$$

By expanding the denominator, substituting this expression in the equation for $P_N(s)$ and then performing the long division, we get

$$P_N(s) = \frac{1}{s^{N+1}} \prod_{i=1}^{N} \lambda_i - \frac{1}{s^N} \prod_{i=1}^{N} \lambda_i \sum_{i=1}^{N} \lambda_i + \cdots \tag{2.37}$$

This equation can be easily inverted using inverse Laplace-Stieltjes transform and we have

$$P_N(t) = \frac{t^N}{N!} \prod_{i=1}^{N} \lambda_i - \frac{t^{N-1}}{(N-1)!} \prod_{i=1}^{N} \lambda_i \sum_{i=1}^{N} \lambda_i + \cdots$$

■

2.4. Nonhomogeneous Poisson Process (NHPP) Models

A *counting process*, $N(t)$, is obtained by counting the number of certain events occurring in the time interval $[0,t)$. The simplest model is the Poisson process model which assumes that time between failures are exponentially distributed and has independent increment, and it has a constant failure occurrence rate over time. Such a model is also a Markov model that has been discussed before. Here we will focus on the case of time-dependent failure occurrence rate, or general NHPP models. Such models are widely used to model the number of failures of a system over time, especially in software reliability analysis (Xie, 1991).

2.4.1. General formulation

Nonhomogeneous Poisson Process (NHPP) models are very useful in reliability analysis, especially for repairable systems. Since hardware systems are usually repairable, and software debugging is a repair process, NHPP models can be used for both software and hardware, and for combined systems.

For a counting process $\{N(t), t \geq 0\}$ modeled by NHPP, $N(t)$ follows a Poisson distribution given the following underlying assumptions of the NHPP:

1) $N(0) = 0$,
2) $\{N(t), t \geq 0\}$ has independent increments,
3) $\Pr\{N(t+h) - N(t) = 1\} = \lambda(t) + o(h)$,
4) $\Pr\{N(t+h) - N(t) \geq 2\} = o(h)$.

In the above $o(h)$ denotes a quantity which tends to zero for small h. The intensity function $\lambda(t)$ is defined as

$$\lambda(t) = \lim_{\Delta t \to 0} \frac{\Pr\{N(t+\Delta t) - N(t) > 0\}}{\Delta t}$$

If we let

$$m(t) = \int_0^t \lambda(t) dt$$

then it can be shown, see e.g. (Ross, 2000: pp. 284-286), that

$$\Pr\{N(t+t_0) - N(t_0) = n\} = \exp\{m(t+t_0) - m(t_0)\} \cdot \frac{[m(t+t_0) - m(t_0)]^n}{n!}, \quad n \geq 0$$

That is, $N(t+t_0) - N(t_0)$ is a Poisson random variable with mean $m(t+t_0) - m(t_0)$. This implies that $N(t)$ is Poisson given $N(0) = 0$ at the initial time $t_0 = 0$, i.e.,

$$\Pr\{N(t) = n\} = \frac{[m(t)]^n}{n!} \exp\{-m(t)\}, \; n=0,1,2,.... \tag{2.38}$$

Here $m(t)$ is called the *mean value function* of the NHPP. If $N(t)$ represents the number of system failures, the function $m(t)$ describes the expected cumulative number of failures in $[0,t)$. Hence, $m(t)$ is a very useful descriptive measure of the failure behavior.

2.4.2. Reliability measures and properties

Given the mean value function $m(t)$, the failure intensity function $\lambda(t)$ can be computed by

$$\lambda(t)=\frac{dm(t)}{dt} \tag{2.39}$$

Moreover, the reliability function at time t_0 is given by

$$R(t\,|\,t_0)=\exp\{m(t_0)-m(t+t_0)\} \tag{2.40}$$

Generally, by using different functions $m(t)$, different NHPP models can be obtained. In the simplest case for which $\lambda(t)$ is constant, the NHPP becomes a homogeneous Poisson process which has a mean value function as t multiplied by a constant.

Similar to the Poisson distribution to which the NHPP is related, it is characterized by several unique and desirable mathematical properties. For example, NHPPs are closed under superposition, that is, the sum of a number of NHPPs is also a NHPP. Generally, we may mix the failure time data from different failure processes assumed to be NHPP and obtain an overall NHPP with a mean value function which is the sum of the mean value functions of the underlying NHPP models.

Any NHPP can be transformed to a homogeneous Poisson process through an appropriate time-transformation. From the general theory of NHPP, it is well-known that if $\{N(t),t\geq 0\}$ is a NHPP with mean value function $m(t)$, then the time-transformed process $N^*(t)$ defined as

$$N^*(t)=N(y(t)),\ \ t\geq 0 \tag{2.41}$$

is also NHPP. The mean value function of the NHPP $\{N^*(t),t\geq 0\}$ is

$$m^*(t) = m(y(t)),\ t \geq 0 \tag{2.42}$$

Especially, if $y(t) = m^{-1}(t)$, we have that the time-transformed process becomes a homogeneous Poisson process with rate one, i.e., the mean value function is equal to t.

Example 2.11. Suppose the mean value function of an NHPP model is

$$m(t) = a \cdot [1 - \exp(-b \cdot t)],\ a > 0, b > 0$$

Let

$$y(t) = m^{-1}(t) = \frac{1}{-b} \ln\left(1 - \frac{t}{a}\right)$$

Then the time-transformed process $N^*(t) = N(y(t))$ is also an NHPP with the mean

$$m^*(t) = m(y(t)) = a \cdot [1 - \exp\{-b \cdot y(t)\}] = t$$

Therefore, the failure intensity function is derived by

$$\lambda^*(t) = \frac{dm^*(t)}{dt} = 1$$

where $\lambda^*(t)$ is a constant which indicates that this time-transformed process becomes a homogeneous Poisson process with constant rate 1.

■

2.4.3. Parameter estimation

Usually, the mean value function $m(t)$ contains some unknown parameters. The estimation of them is generally carried out by using the method of maximum likelihood or the method of least squares.

Denote by n_i the number of faults detected in the time interval $[t_{i-1}, t_i)$, where $0 = t_0 < t_1 < \cdots < t_k$ and t_i is the running time since the beginning. The likelihood function for the NHPP model with mean value function $m(t)$ is

$$L(n_1, n_2, \ldots n_k) = \prod_{i=1}^{k} \frac{[m(t_{i-1}) - m(t_i)]^{n_i} \exp\{m(t_{i-1}) - m(t_i)\}}{n!}.$$

The parameters in $m(t)$ can then be estimated by maximizing this likelihood function. Usually, numerical procedures have to be used in solving the likelihood equations.

CHAPTER 3

MODELS FOR HARDWARE SYSTEM RELIABILITY

In the computing systems, hardware (such as hard disk, router, processor, CPU, memory, etc.) provides the fundamental configurations to support computing tasks. This chapter focuses on the methods and models that are commonly used in analyzing the hardware reliability. They are also useful for integrated systems which will be discussed in later chapters.

Reliability models for single component system are first presented. Then, some models of parallel configurations are studied. Following that, some other techniques in fault tolerance system including load-sharing and standby configurations are also shown.

3.1. Single Component System

We first consider a system with one component or when the system is considered as a black-box. A single hardware component may have a normal functioning state, a few degraded states and a failed state. This section analyzes the reliability performance of the single component, considering a single failure mode, double failure modes and multiple failure modes.

3.1.1. Case of a single failure mode

Suppose that there are two states, and a single, irreversible transition between the two states as shown in Fig. 3.1. The two states are operational state and failed state denoted by state 1 and 2 respectively. Such a case is called single failure mode case here.

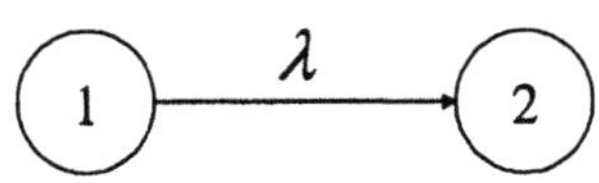

Fig. 3.1. State transition diagram for a single component with a single failure mode.

In Fig. 3.1, λ is the transition rate from state 1 to state 2, and it corresponds to the failure rate of the hardware component whose lifetime is assumed to follow exponential distribution. The component reliability (the probability of being in state 1) is given by

$$R(t) = P_1(t) = \exp(-\lambda t) \tag{3.1}$$

If the component is repairable with the repair rate μ, the Markov model is shown by Fig. 3.2.

Fig. 3.2. State transition diagram for a repairable component.

The reliability function that the component first reaches the failed state is also $\exp(-\lambda t)$. However, the system availability function is the probability for the component to stay at operational state (state 1) at the time instant t and it is given by

$$A(t) = P_1(t) \tag{3.2}$$

To obtain $P_1(t)$, the Chapman-Kolmogorov equations can be written as

$$P_1'(t) = \mu P_2(t) - \lambda P_1(t) \tag{3.3}$$

$$P_2'(t) = \lambda P_1(t) - \mu P_2(t) \tag{3.4}$$

Since the system has to be at state 1 or state 2, $P_2(t) = 1 - P_1(t)$. Substituting this into the above equations, we get

$$P_1'(t) = -(\mu + \lambda)P_1(t) + \mu \tag{3.5}$$

With the initial conditions

$$P_1(0) = 1,\ P_2(0) = 0 \tag{3.6}$$

we can obtain the availability function as

$$A(t) = P_1(t) = \frac{\lambda}{\mu + \lambda}\exp\{-(\mu + \lambda)t\} + \frac{\mu}{\mu + \lambda} \tag{3.7}$$

Example 3.1. Suppose that a hardware system has been working for 1000 hours during which the system failed 30 times and the total repair time for all the failures is 150 hours. If the hardware failure time and repair time follow the exponential distributions, then the expected failure rate and repair rate can be estimated by

$$\lambda = \frac{30}{1000} = 0.03 \text{ and } \mu = \frac{30}{150} = 0.2 \tag{3.8}$$

The reliability function is

$$R(t) = \exp(-0.03t)$$

and the availability function is

$$A(t) = P_1(t) = 0.1304\exp(-0.23t) + 0.8696$$

The curve for availability function $A(t)$ is depicted by Fig. 3.3.

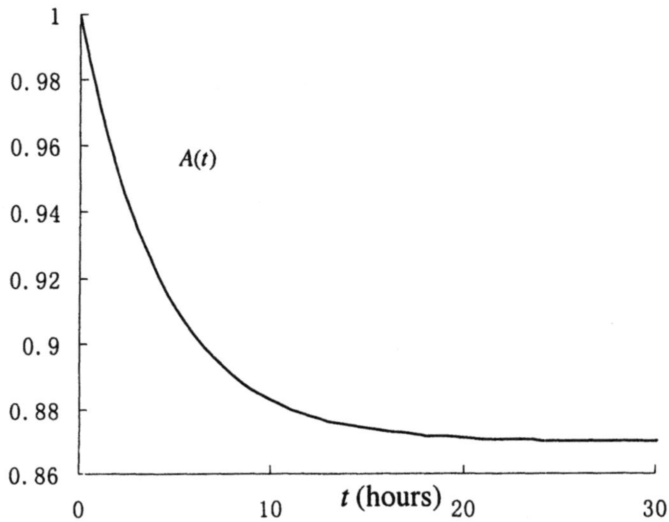

Fig. 3.3. The availability function for Example 3.1.

■

3.1.2. Case of double failure modes

The reliability evaluation above is based on faults which are permanent in nature. By considering the *double failure modes* including both intermittent failures and permanent failures (Prasad, 1991), a reliability model is presented here. The hardware component is assumed to start from an operational state and can go to either an intermittent failure state or a permanent failure state. The intermittent failure can also lead the component into the permanent failure state. This scenario is presented by a Markov model as shown in Fig. 3.4.

In Fig. 3.4, the states 0, 1 and 2 are operational, intermittent failure and permanent failure states, respectively. According to the model, the state 0 can make a transition to state 1 with a rate v and to state 2 with a rate λ. From the intermittent failure state 1, it can either go to operational state 0 with a rate μ or to permanent failure state 2 with a rate λ.

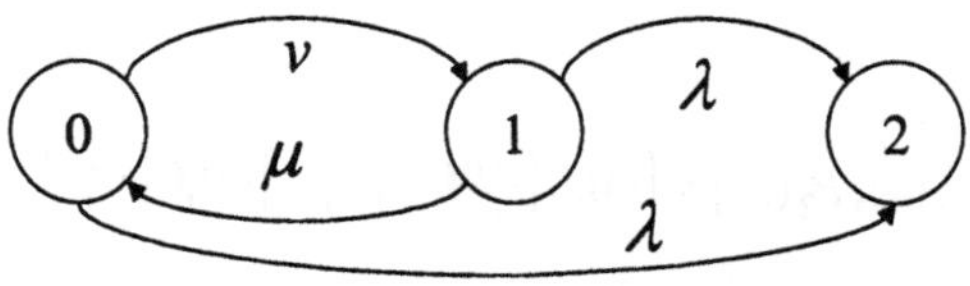

Fig. 3.4. Markov model considering intermittent faults.

Let $P_0(t)$, $P_1(t)$ and $P_2(t)$ be the probabilities of being in the states 0, 1 and 2, respectively. From Fig. 3.4, a set of Chapman-Kolmogorov equations can be written in the matrix form as:

$$[P_0'(t), P_1'(t), P_2'(t)] = [P_0(t), P_1(t), P_2(t)] \cdot T \tag{3.9}$$

where the transition matrix T is given by

$$T = \begin{bmatrix} -\nu-\lambda & \nu & \lambda \\ \mu & -\mu-\lambda & \lambda \\ 0 & 0 & 0 \end{bmatrix} \tag{3.10}$$

Taking the Laplace-Stieltjes transform, we get

$$[P_0(0), P_1(0), P_2(0)] = [P_0(s), P_1(s), P_2(s)] \cdot [S\mathbf{I} - T] \tag{3.11}$$

where

$$[S\mathbf{I} - T] = \begin{bmatrix} S+\nu+\lambda & -\nu & -\lambda \\ -\mu & S+\mu+\lambda & -\lambda \\ 0 & 0 & S \end{bmatrix} \tag{3.12}$$

and $P_i(0)$ is the initial value of $P_i(t)$ at t=0, i=0,1,2. Hence, we have

$$[P_0(s), P_1(s), P_2(s)] = [P_0(0), P_1(0), P_2(0)] \cdot [S\mathbf{I} - T]^{-1} \tag{3.13}$$

Assuming that the system starts from the operational state, then the boundary condition is

$$[P_0(0), P_1(0), P_2(0)] = [1,0,0].$$

Hence,

$$[P_0(s), P_1(s), P_2(s)] = [1,0,0]\cdot[s\mathbf{I} - T] = [1,0,0]\cdot A^{-1} \tag{3.14}$$

and the inverse of the matrix is given by:

$$A^{-1} = \frac{1}{\det A}\begin{bmatrix} s^2 + s(v+\mu) & sv & s\lambda + \lambda(v+\mu+\lambda) \\ s\mu & s^2 + s(v+\lambda) & s\lambda + \lambda(v+\mu+\lambda) \\ 0 & 0 & s^2 + s(v+\mu+2\lambda) + \lambda(v+\mu+\lambda) \end{bmatrix} \tag{3.15}$$

where

$$\det A = s^3 + s^2(v+\mu+2\lambda) + s(\lambda v + \lambda\mu + \lambda^2). \tag{3.16}$$

Solving for $P_0(s)$, we obtain (Prasad, 1991),

$$P_0(s) = \frac{s+\lambda+\mu}{s^2 + s(2\lambda+v+\mu) + \lambda v + \lambda\mu + \lambda^2} \tag{3.17}$$

Taking the inverse Laplace-Stieltjes transform, the system availability function can be obtained as

$$A(t) = P_0(t) = \frac{\mu}{v+\mu}\exp(-\lambda t) + \frac{v}{v+\mu}\exp\{-(\lambda+v+\mu)t\} \tag{3.18}$$

Example 3.2. Suppose that for a computing system, the rate for intermittent failures to occur is $v = 0.02$ and for permanent failures $\lambda = 0.01$. The repair rate from the intermittent failure state to operational state $\mu = 0.08$. Substitute them into the above availability function, we get

$$\begin{aligned} A(t) = P_0(t) &= \frac{\mu}{v+\mu}\exp(-\lambda t) + \frac{v}{v+\mu}\exp\{-(\lambda+v+\mu)t\} \\ &= 0.8\exp(-0.01t) + 0.2\exp(-0.11t) \end{aligned}$$

■

3.1.3. Case of multiple failure modes

This model with multiple failure modes applies if the given component can fail in several modes. These modes have different effects on the system operations, e.g. Levitin *et al.* (1998). The Markov diagram for a component with three failure modes, such as a component that can fail in either open or shorted mode or may experience drift outside the specified range, has the following states:

State 1: Component is fully operational.

State 2: Component has failed in open mode.

State 3: Component has failed in shorted mode.

State 4: Component has drifted outside specification values.

Note that in this case states 2 and 3 will be failed states and state 4 a degraded state. The Markov transition diagram for this case is shown in Fig. 3.5.

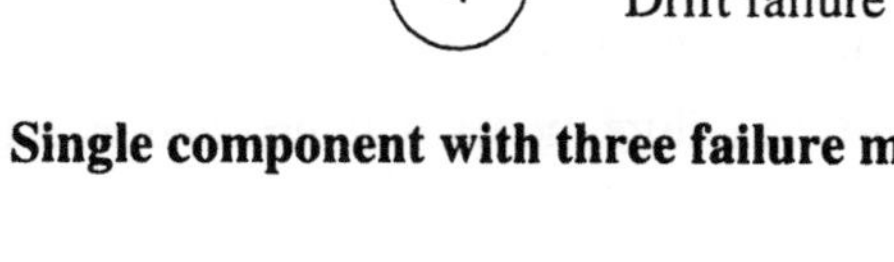

Fig. 3.5. Single component with three failure modes.

In effect, the total failure rate for the component is given by

$$\lambda_T = \lambda_D + \lambda_o + \lambda_s$$

Generally, if the hardware component has n-type of failure modes and the transition among different failure modes is allowed, the Markov transition diagram is depicted by Fig. 3.6.

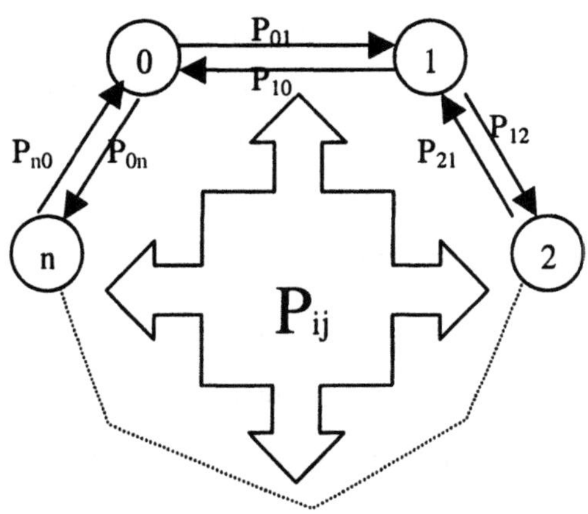

Fig. 3.6. Markov diagram for *n*-type failure modes.

In Fig. 3.6, the state 0 is the fully operational state, and states 1 to n represent the n different failure modes. Denote the transition probability from state i to state j by

$$P_{ij} = P\{Z_{k+1} = j \mid Z_k = i\}, \quad i, j = 0,1,2,\cdots,n$$

In fact, the models of single failure mode and double failure modes are two special cases of the n-type failure modes with n=1 and n=2, respectively.

3.2. Parallel Configurations

Parallel system is one of the most frequently used redundancy configurations in order to achieve fault-tolerance which is important in computing systems. A parallel configuration assumes that the failure of a component will not affect the operation of the remaining components and all the components can support the functions of one another.

3.2.1. Two-component parallel configuration

As the simplest parallel system, two-component configuration is studied here. Its reliability block diagram is shown in Fig. 3.7.

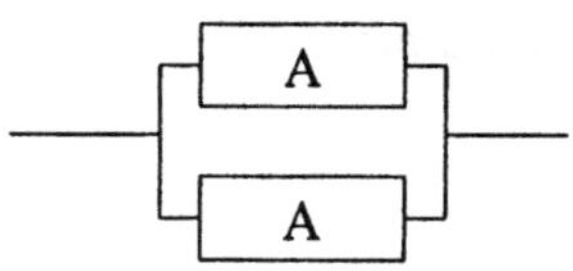

Fig. 3.7. Two-component parallel configuration reliability block diagram.

For this two-component parallel configuration, if both components are identical, there are three states:

State 1: Two components are operational.

State 2: Only one component is operational.

State 3: System has failed (all components have failed).

The applicable Markov transition diagram for the parallel two-component redundant system is depicted by Fig. 3.8.

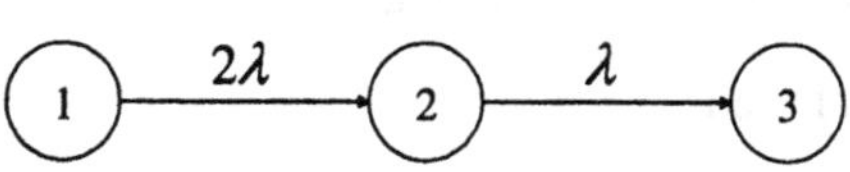

Fig. 3.8. Markov model for two parallel components.

The solution for the system reliability can be shown to be

$$R(t) = 1 - P_3(t) = 1 - [1 - \exp(-\lambda t)]^2$$

3.2.2. Majority voting configuration

Majority voting systems form an important class of redundant systems. In a majority voting system, all of the components are assumed to be in operation. Many voting systems for the *N*-component hardware are based on the majority rule, see e.g., Ashrafi *et al.* (1994).

The simplest majority voting system consists of three components and a voter. The reliability block diagram for a majority voter configuration is shown in Fig. 3.9.

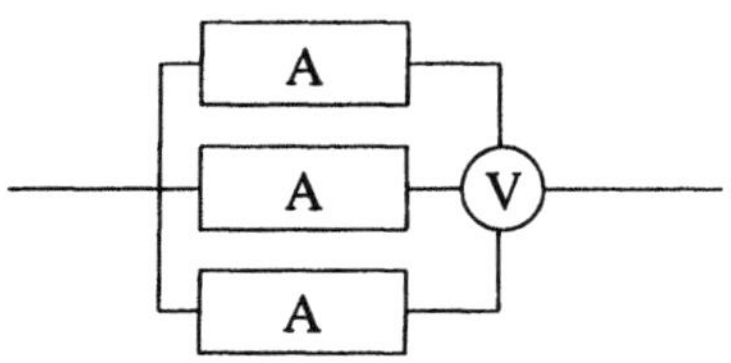

Fig. 3.9. Triple majority voter reliability block diagram.

This configuration is also known as a triple modular redundancy configuration, and it requires at least two good components for operation. Assuming that the voter is perfect, the system states are:

State 1: Three components are operational.

State 2: Two components are operational.

State 3: System has failed.

The Markov transition diagram is shown in Fig. 3.10. The solution for evaluating the reliability of the triple modular redundancy configuration is

$$R(t) = \exp(-3\lambda t) + 3\exp(-2\lambda t)[1 - \exp(-\lambda t)]$$

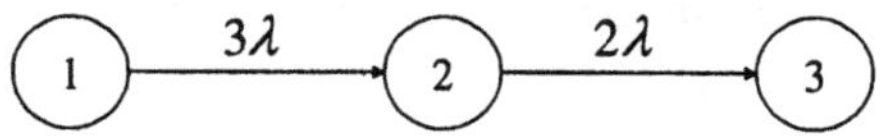

Fig. 3.10. State transition diagram for three-component majority voting (Triplex-duplex).

A modified scheme

It is also possible to increase the reliability of the triple modular redundancy system by making a simple modification in the operating sequence. After the first failure has been detected, there are two remaining modules or components. There is usually no need to keep both of the remaining components, since it will not be possible to identify the failed component after the second failure. The resulting state transition diagram will become as shown in Fig. 3.11.

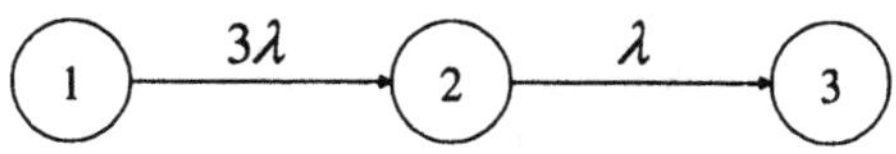

Fig. 3.11. State transition diagram for three-component majority voting (Triplex-simplex).

The reliability function can be derived as

$$R(t) = \exp(-3\lambda t) + 3\exp(-\lambda t)[1 - \exp(-\lambda t)]$$

Example 3.3. A three-component majority voting system has the failure rate $\lambda = 0.01$ for each parallel component.

Without removing any component when the first component fails, the reliability function is

$$R_1(t) = \exp(-0.03t) + 3\exp(-0.02t)[1 - \exp(-0.01t)]$$

The probability for the system to successfully complete a 10-hour mission can be computed as $R_1(10) = 0.9746$.

If we remove either one component if the first component fails, the reliability function is

$$R_2(t) = \exp(-0.03t) + 3\exp(-0.01t)[1 - \exp(-0.01t)]$$

The probability for the system to work well in 10 hours is $R_2(10) = 0.9991$. Both curves for the two reliability functions $R_1(t)$ and $R_2(t)$ are depicted by Fig. 3.12.

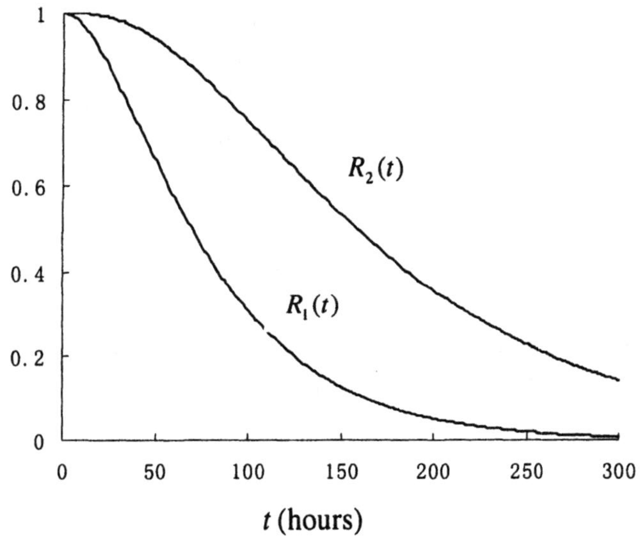

Fig. 3.12. Two reliability functions curves for the Example 3.3.

■

3.2.3. *k*-out-of-*N* voting configuration

This redundancy configuration is known as the *N*-Modular Redundancy. The configuration requires that *k* functional components out of a total of *N* are needed for the system to remain operational. Akhtar (1994) presented a Markov model to analyze the k-out-of-N voting system for both perfect and imperfect fault-coverage problems. The failures in the system may be covered or uncovered. Fault coverage is a measure of the ability to perform fault detection, fault location, fault containment, or fault recovery.

Perfect fault-coverage modeling

Fig. 3.13 shows the Markov chain for the N-component system with perfect fault-coverage. The process is birth-death process with a constant failure rate, denoted by λ for each component. Here μ_i is repair rate for state i, $0 < i \leq N$.

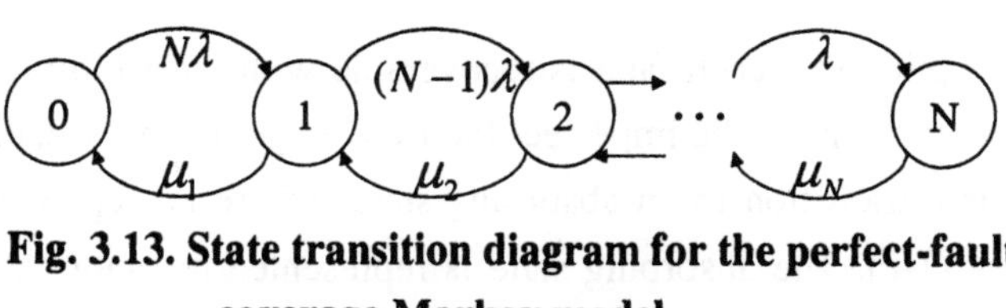

Fig. 3.13. State transition diagram for the perfect-fault coverage Markov model.

State i means i components have failed and the rest are operational. The probability for staying at the i:th state is denoted by $P_i(t)$, which can be easily obtained by solving the following Chapman-Kolmogorov equation:

$$P_0'(t) = \mu_1 P_1(t) - N\lambda P_0(t)$$

$$P_i'(t) = (N - i + 1)\lambda P_{i-1}(t) + \mu_{i+1} P_{i+1}(t) - [(N - i)\lambda + \mu_i]P_i(t), \quad i = 1,2,3,...N-1$$

$$P_N'(t) = \lambda P_{N-1}(t) - \mu_1 P_N(t)$$

with the initial conditions

$$P_0(0) = 1, P_i(0) = 0, \quad i=1,2,\ldots,N$$

The system availability can be computed through

$$A(t) = \sum_{i=0}^{N-k} P_i(t) \tag{3.19}$$

and the system reliability can be obtained by considering a pure birth process with $\mu = 0$ through the equation:

$$R(t) = \sum_{i=0}^{N-k} P_i(t) \tag{3.20}$$

Since there is no absorbing state in Fig. 3.13, the steady-state availability can be calculated by

$$A = \sum_{i=0}^{N-k} P_0 \frac{(N-i)\lambda}{\mu_{i+1}} \frac{k!}{(k-i)!} \tag{3.21}$$

Imperfect fault-coverage modeling

Under the assumption that each fault is recoverable with probability c, Fig. 3.14 shows the Markov chain for the imperfect fault-coverage model, see, e.g., Akhtar (1994). There is a transition to an absorbing state (where no repair is possible) with probability (1-c). The absorbing state is represented by state "N+1". Thus, there are N+2 states, denoted by $\Omega_c = \{0,1,...,N,N+1\}$.

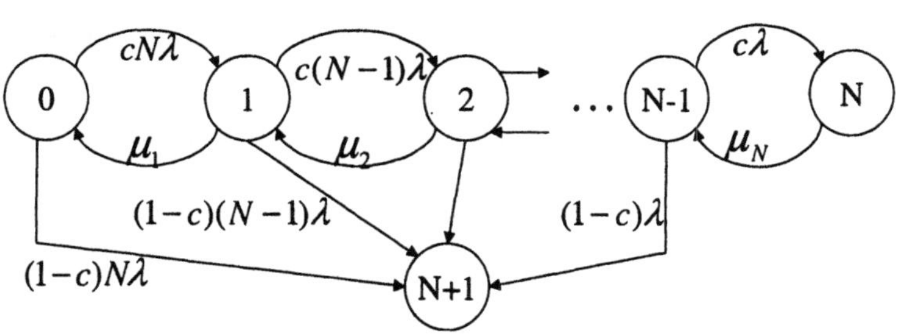

Fig. 3.14. State transition diagram for the imperfect fault coverage Markov model.

As Fig. 3.14, the model is obtained by considering 3 classes of states:

State 0: all units are operational.

State i ($1 \le i \le N$): i of the components have failed with repair possible at all i states, and the system transits from state i to i-1 with rate μ_i ($\mu_i = \mu$ for a single-repair facility; $\mu_i = i \cdot \mu$ for a multiple-repair facility).

State N+1: a system failure state, where repair is not possible.

The Chapman-Kolmogorov equations can be given by

$$\begin{aligned} P_i'(t) &= (N-i+1)\lambda c P_{i-1}(t) \\ &\quad -[(N-i)\lambda + \mu_i]P_i(t) + \mu_{i+1}P_{i+1}(t) \end{aligned} \qquad 1 \le i \le N-1 \tag{3.22}$$

$$P_0'(t) = -N\lambda \cdot P_0(t) + \mu_1 P_1(t) \tag{3.23}$$

$$P_N'(t) = \lambda c \cdot P_{N-1}(t) - \mu_N P_N(t) \tag{3.24}$$

and

$$P_{N+1}'(t) = \sum_{i=0}^{N-1} (N-i)\lambda(1-c) \cdot P_i(t) \tag{3.25}$$

Denote by $X = S\mathbf{I} - Q$ where Q is the transition-rate matrix by excluding the last row and last column of the whole transition-rate matrix. With the initial condition $[P_0(0), P_1(0), ..., P_N(0)] = [1,0,0,...,0]$, Akhtar (1994) showed that

$$P_i(t) = \sum_{j=1}^{N+1} \left[\frac{D_i(s)|_{s=-r_j} \exp(-r_j \cdot t)}{\prod_{z=1, z \ne j}^{N+1} (r_z - r_j)} \right] \quad \text{for } 0 \le i \le N \tag{3.26}$$

and

$$P_{N+1}(t) = 1 - \sum_{i=0}^{N} P_i(t) \tag{3.27}$$

where r_i ($0 \le i \le N+1$) are the roots of $|X| = 0$, and $D_i(s)$ is the determination of the matrix that replaces the i:th column of X by the initial vector $[P_0(0), P_1(0), ..., P_N(0)]^T$.

The system availability can be computed with

$$A(t) = \sum_{i=0}^{N-k} P_i(t) \tag{3.28}$$

The system reliability $R(t)$ can be derived by considering a pure birth process as

$$R(t) = \sum_{i=0}^{N-k} P_i(t),\ \mu_i = 0 \tag{3.29}$$

The perfect fault-coverage model is a special case of the imperfect model by fixing c=1. The above results of availability and reliability functions can be similarly implemented in the perfect case by substituting c=1 in those equations.

Example 3.4. Consider a 1-out-of-3 system with imperfect fault-coverage. Suppose $\mu_i = i \cdot \mu$ for a multiple-repair facility and the numerical values for $\lambda = 10^{-6}$, $\mu = 10^{-5}$ and $c = 0.95$. The Markov model will contain five states $\{0,1,2,3,4\}$ as the Fig. 3.14, where state 4 is an absorbing failure state, and state 3 is a non-absorbing failure state.

For availability function $A(t)$, from the state transition rate, we have

$$|X| = \begin{vmatrix} s+3\lambda & -\mu & 0 & 0 \\ -3\lambda c & s+2\lambda+\mu & -2\mu & 0 \\ 0 & -2\lambda c & s+\lambda+2\mu & -3\mu \\ 0 & 0 & -\lambda c & s+3\mu \end{vmatrix} \tag{3.30}$$

The four real roots are obtained by solving $|X|$=0 using numerical method:

r_1=1.36932, r_2=110.456, r_3=219.544, r_4=328.631,

Finally, the state probabilities are obtained as

$$P_i(t) = \sum_{j=1}^{4} \left[\frac{D_i(s)\,|_{s=-r_j} \exp(-r_j \cdot t)}{\prod_{z=1, z\neq j}^{4} (r_z - r_j)} \right] \quad \text{for } 0 \le i \le 3 \tag{3.31}$$

where

$$D_0(s) = (s + 2\lambda + \mu)\cdot[(s + \lambda + 2\mu)\cdot(s + 3\mu) - 3\lambda\cdot c\cdot\mu] - 4\lambda\cdot c\cdot\mu\cdot(s + 3\mu)$$

$$D_1(s) = 3\lambda\cdot c\cdot[(s + \lambda + 2\mu)\cdot(s + 3\mu) - 3\lambda\cdot c\cdot\mu]$$

$$D_2(s) = 6(\lambda\cdot c)^2\cdot(s + 3\mu)$$

$$D_3(s) = -6(\lambda\cdot c)^3$$

and

$$P_4(t) = 1 - P_0(t) - P_1(t) - P_2(t) - P_3(t) \tag{3.32}$$

The numerical results of the five state probabilities are plotted in Fig. 3.15, where $P_2(t)$ and $P_3(t)$ are almost 0.

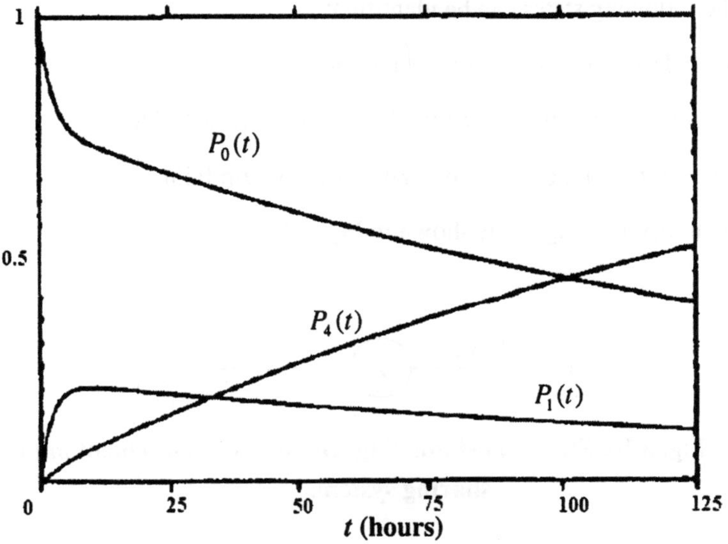

Fig. 3.15. State probabilities for Example 3.4.

The availability and the reliability functions can be obtained accordingly using Eqs. (3.28) and (3.29).

■

3.3. Load-Sharing Configurations

While components in parallel systems are designed to carry full load, in load-sharing systems, each component is designed to carry only part of the load. If one component fails in the load-sharing system, then the remaining components share its load. Furthermore, since the components now carry heavier load, their failure rates will increase due to the additional stress.

3.3.1. Two-component load-sharing system

Consider a parallel load-sharing system consisting of two components. Under the load-sharing conditions, assume each component carries only one-half of the load. The following states can be identified:

State 1: Two components are operational on a load-sharing basis.

State 2: One component has failed, the other carries full load.

State 3: Both components have failed, i.e., system failure.

The state transition diagram is shown in Fig. 3.16.

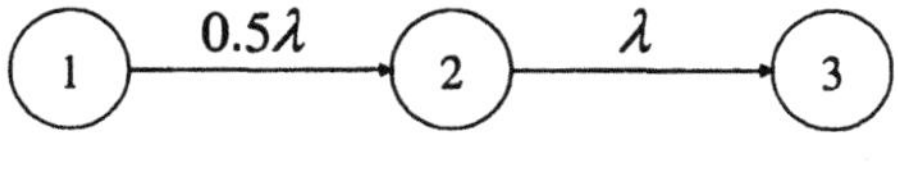

Fig. 3.16. State transition diagram for two-component load-sharing system.

Here, the transition rate for the first transition is only one-half that for the full-load parallel system. The system reliability function is then given by

$$R(t) = \exp(-0.5\lambda t) + 2[1 - \exp(-0.5\lambda t)]\exp(-\lambda t)$$

3.3.2. *k*-out-of-*N* load-sharing system

A load-sharing *k*-out-of-*N* system is a configuration which works if at least *k* out of *N* components are functioning and the surviving subsystems share the total load. Shao & Lamberson (1991) studied such a system. The assumptions of the model are given as follows:

1) The failure rate of all functioning components is the same and a functioning unit of *i* components has the constant failure rate, λ_i, $i = k,..., N$.

2) A failed component must be detected and disconnected by a controller and the probability of success is α. If the controller cannot detect and disconnect a failed unit or the controller itself has failed, the system fails. The controller failure rate is a constant, denoted by λ_c.

3) At most *r* components can be in repair at one time each with a repair rate μ, so the repair rate for *j* components failed is: $\mu_j = \mu \min(j, r)$.

4) A repaired component is as good as new and is immediately reconnected to the system with negligible switch-over time.

5) The controller is never repaired or replaced during a mission.

Based on the above assumptions, the Markov model can be constructed as depicted by Fig. 3.17.

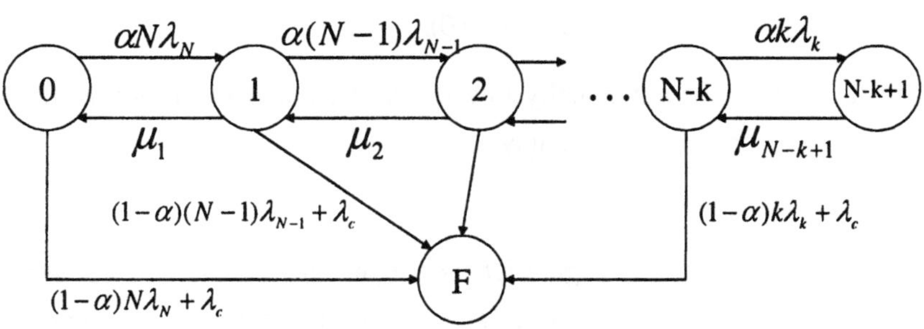

Fig. 3.17. Markov transition diagram for load-sharing *k*-out-of-*N* system.

As shown in Fig. 3.17, the state space for the system is defined below:

State j (j=0,1,...,N-k): j components have failed and have been disconnected from the network, the remaining (N-j) components and the controller are functioning.

State N-k+1: the system fails because only (k-1) components are functioning, but the system can return to working state (N-k) at a repair rate μ_{N-k+1}.

State F: the system fails because the controller cannot detect and disconnect a failed unit.

The Chapman-Kolmogorov equations can be given by

$$P_0'(t) = -(N\lambda_N + \lambda_c)P_0(t) + \mu_1 P_1(t)$$

$$\begin{aligned} P_j'(t) &= -[(N-j)\lambda_{N-j} + \lambda_c + \mu_j]P_j(t) \\ &\quad + \alpha(N-j+1)\lambda_{N-j+1}P_{j-1}(t) + \mu_{j+1}P_{j+1}(t), \end{aligned} \qquad j = 1,2,\ldots,N-k$$

$$P_{N-k+1}'(t) = -\mu_{N-k+1}P_{N-k+1}(t) + \alpha k\lambda_k P_{N-k}(t) \tag{3.33}$$

$$P_F'(t) = (1-\alpha)\sum_{j=0}^{N-k}[(N-j)\lambda_{N-j}P_j(t) + \lambda_c]$$

The initial conditions are:

$$P_i(0) = \begin{cases} 1, & (i = 0) \\ 0, & (i \neq 0) \end{cases} \tag{3.34}$$

These equations can be numerically solved. The system availability function and reliability can be obtained accordingly.

Example 3.5. Consider a jet engine functioning under full load on a commercial airplane. Two functioning jet engines are required for flying, but 4 engines are functioning for full power. An engine controller manages the load-sharing. When

4 engines function in the airplane, the load on each is much less than when they function alone. From the test data, if 4 engines are functioning for an airplane, the failure rate for each engine is reduced to 50%, while if three engines are functioning, the failure rate is reduced to 60% and two engines to 70%. The switching probability, α =0.99, the jet engine failure rate under the full load λ=0.1, $\lambda_c = 0.01$ and repair rate $\mu_i = 0.2$ (*i*=1,2,3).

The above jet engine system is a 2-out-of-4 load-sharing system. Its CTMC can be modeled as Fig. 3.18.

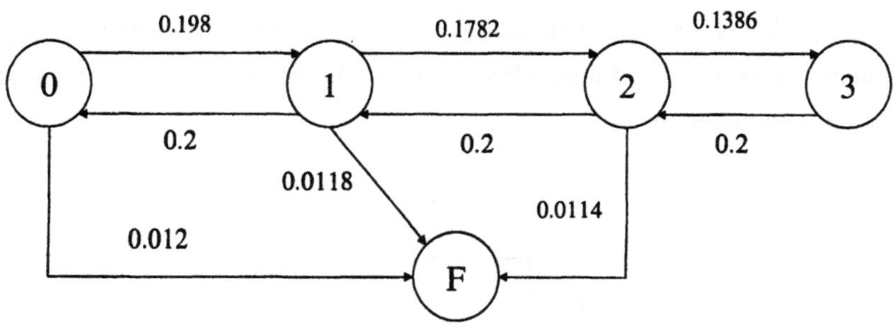

Fig. 3.18. Markov model for the 2-out-of-4 load-sharing system.

The Chapman-Kolmogorov equation can be constructed. It is then possible to solve for all the state probability functions $P_i(t)$ and then obtain the system availability function $A(t)$.

■

3.4. Standby Configurations

Standby redundancy is particularly important in those applications where low power consumption is mandatory, such as in spacecraft systems. Standby systems also yield better reliability than can be achieved using the same quantity of equipment in parallel mode. This happens when the standby condition failure rate is assumed to be zero. If this assumption does not apply, the model needs to be modified to account for the storage failure rate. Moreover, the switch and monitor to control the system may fail caused by their own faults, which will also be considered in this section. Finally, the multi-mode operations for the standby redundancies will be discussed as well.

3.4.1. Standby with zero storage failure rate

Usually standby components can be assumed to have zero or very low failure rate in storage. If this is the case, then we have a simple system consisting of only two components, a primary and a standby spare, as shown in Fig. 3.19. The spare is passive until switched in.

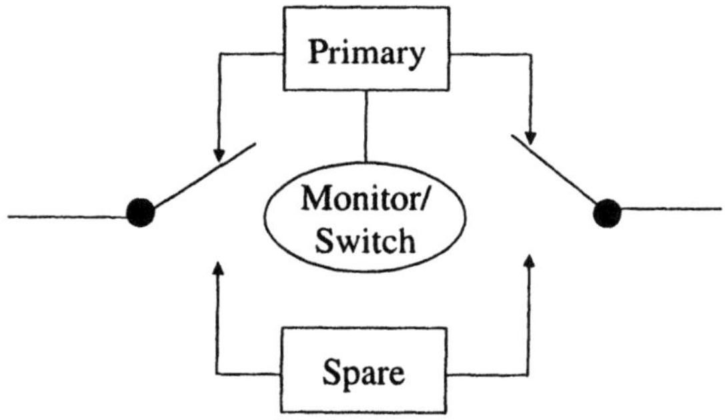

Fig. 3.19. Standby configuration.

Both components are assumed to have the same failure rate, λ, when operating. In the standby mode, the failure rate is zero (i.e. cold standby). Since only one of these components is used at a given time, we identify the following states:

State 1: Primary component is operational.

State 2: Standby component has been switched in and is operational.

State 3: Both components have failed, i.e., system failure.

The transition rate from state 1 to state 2 and that from state 2 to state 3 are equal to λ. The reliability function can be obtained as

$$R(t) = \exp(-\lambda t) + [1 - \exp(-\lambda t)]\exp(-\lambda t)$$

The same approach can be extended to standby systems where there are (N-1) cold standby components together with one primary component. The state transition diagram is depicted by Fig. 3.20, where state N+1 is the system failure state.

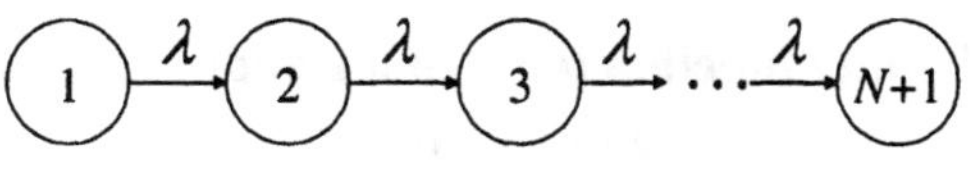

Fig. 3.20. CTMC for *N*-1 cold standby components together with one primary.

The reliability function can be obtained as

$$R(t) = \sum_{i=1}^{N} [1 - \exp(-\lambda t)]^{i-1} \exp(-\lambda t) \tag{3.35}$$

Example 3.6. A system contains two cold standby components and one primary component each of which has the failure rate $\lambda = 0.02$. Then, its reliability function is computed as

$$R(t) = e^{-0.02t} + [1 - e^{-0.02t}]e^{-0.02t} + [1 - e^{-0.02t}]^2 e^{-0.02t}$$

and the curve is shown in Fig. 3.21.

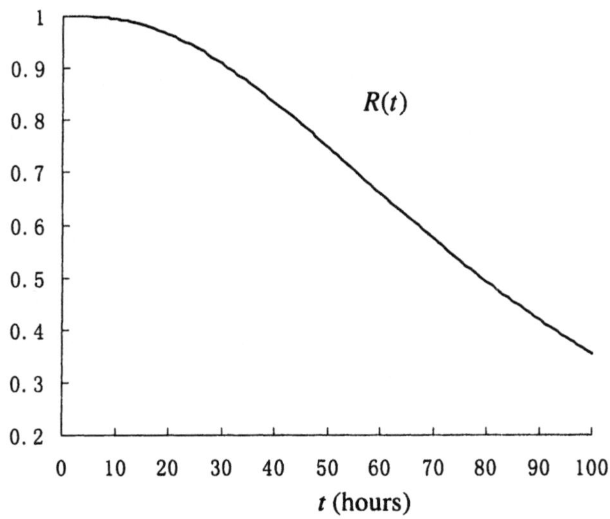

Fig. 3.21. Reliability function for 2-cold standby and one primary component.

■

3.4.2. Standby with nonzero storage failure rate

If the standby component has a nonzero failure rate of λ_s (such as it is energized as a warm or hot component), see e.g. Pukite & Pukite (1998, pp. 73-80), then we can identify these states:

State 1: Both components are good, primary component is operating.

State 2: Primary component has failed; secondary has been switched on and is operational.

State 3: Standby component has failed; system is still operating with primary component.

State 4: Both components have failed; system failure.

The state transition diagram is depicted by Fig. 3.22.

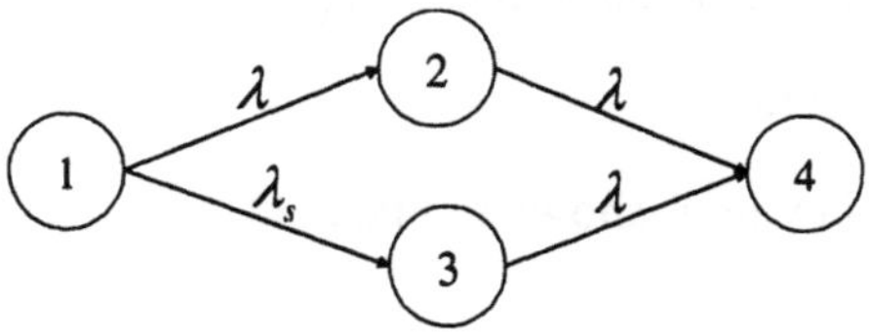

Fig. 3.22. Standby with nonzero storage failure rate.

After merging states 3 and 2 P_3 into P_2, we can easily obtain the reliability function as

$$R(t) = \exp\{-(\lambda + \lambda_s)t\} + [1 - \exp\{-(\lambda + \lambda_s)t\}]\exp(-\lambda t)$$

3.4.3. Imperfect monitor and switch

In the models described so far, the monitors and switches are assumed to be perfectly operational. In this section, we include the effects of imperfect monitor and switch, i.e. we consider the failure of the fault monitor and switch, respectively.

The conventional failure monitor and switch can fail in one of two modes:

1. In a state where the failure monitoring ability is disabled.
2. In a state where a false switching to the next standby component has occurred.

If we assume equal failure rates to the primary and secondary components and initially ignore the component storage failure rates, then by assigning λ_{s1} and λ_{s2} to the monitor and switch failure rates for the two modes described above, the system states will be:

State 1: Primary component, fault monitor, and switch are in operational condition.

State 2: Primary component is operating, but the switch has failed.

State 3: Secondary component is operating.

State 4: System failure.

The state transition diagram is shown in Fig. 3.23.

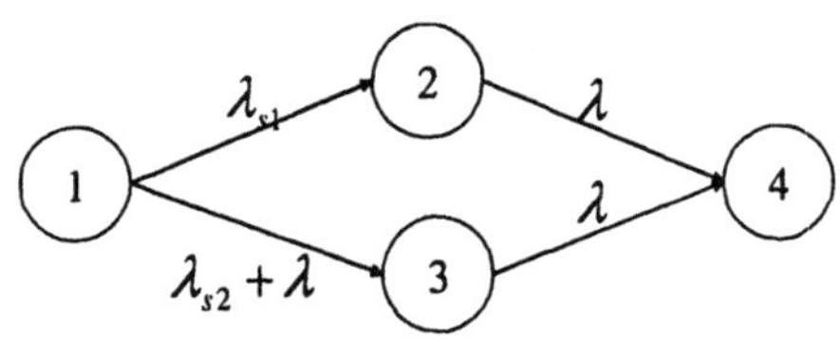

Fig. 3.23. Standby configuration with imperfect monitor and switch.

Note that the states 2 and 3 are identical, in which there is only one operational component left. We can reduce the number of system states and simplify the state diagram. The final reliability function is given by

$$R(t) = \exp\{-(\lambda + \lambda_{s1} + \lambda_{s2})t\} + [1 - \exp\{-(\lambda + \lambda_{s1} + \lambda_{s2})t\}]\exp(-\lambda t)$$

It is possible to extend the same concept to system configurations with more than one standby component by viewing the imperfect monitor/switch as another parallel component.

3.4.4. Multiple mode operation

Many standby systems are designed for multiple mode operation. Chen & Bastani (1992) constructed a CTMC to evaluate the reliability of multiple mode operation system with both full and partial redundancies. The assumptions in their model are given below:

1) Failure times of components are exponentially distributed with a constant failure rate.

2) A full redundancy requires the full power of a component and serves as either a primary component or a hot standby of that component.

3) A partial redundancy requires part of the processing power of the full primary component and serves as a warm standby of that component.

Suppose a system has a full redundancy or primary with failure rate λ_1, and a partial redundancy with failure rate λ_2. Using the Markov model for this system, the reliability function can be obtained as

$$R(t) = \exp(-\lambda_1 t) + [1 - \exp(-\lambda_1 t)]\exp(-\lambda_2 t) \tag{3.36}$$

Also, this Markov model can be extended to an N modes operation system. Suppose a system has one primary component with failure rate λ_1 and N-1 partial redundancies with failure rates (λ_2,..., λ_N). The reliability function can be obtained as

$$R(t) = \exp(-\lambda_1 t) + \sum_{i=2}^{N}\left\{\exp(-\lambda_i t)\cdot\prod_{j=1}^{i-1}[1 - \exp(-\lambda_j t)]\right\} \tag{3.37}$$

Example 3.7. Suppose a system contains one primary with failure rate 0.03 and two partial redundancies with same failure rate of 0.05. Substitute the $\lambda_1 = 0.03$ and $\lambda_2 = \lambda_3 = 0.05$ into the Eq. (3.37), we have

$$R(t) = \exp(-0.03t) + [1 - \exp(-0.03t)]\exp(-0.05t)$$
$$+ [1 - \exp(-0.03t)][1 - \exp(-0.05t)]\exp(-0.05t)$$

The curve of the reliability function is shown in Fig. 3.24.

If we further consider a system with two partial redundancies having nonzero storage failure rate, say 0.01, the Markov model is constructed as the CTMC in Fig. 3.25.

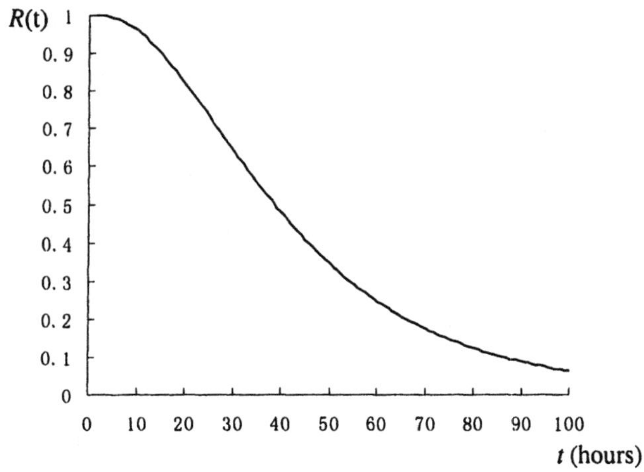

Fig. 3.24. The reliability function in Example 3.7.

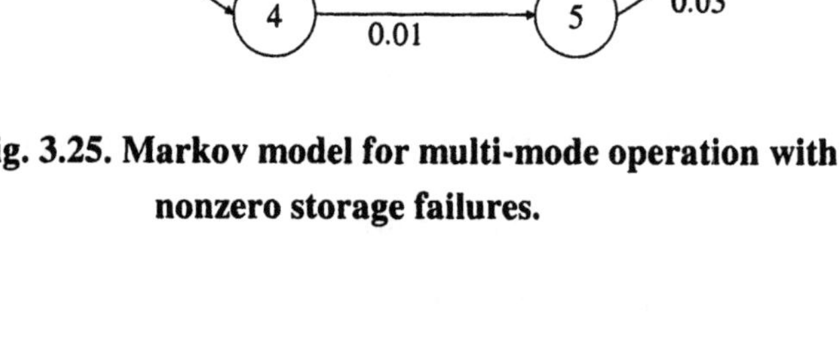

Fig. 3.25. Markov model for multi-mode operation with nonzero storage failures.

At state 1, all the three components are functioning. At state 2, the primary fails and the other two are functioning. At state 3, one primary and one redundancy fail while the other is functioning. At state 4, the primary and one redundancy are functioning but the other redundancy fails due to nonzero storage failure. At state 5, the primary is functioning while the two redundancies fail due to nonzero storage failure. Finally, at state F, all the three components have failed and the system fails.

The Chapman-Kolmogorov equations can be written as

$$P_1'(t) = -0.05P_1(t)$$

$$P_2'(t) = 0.03P_1(t) - 0.06P_2(t)$$

$$P_3'(t) = 0.06P_2(t) + 0.03P_4(t) - 0.05P_3(t)$$

$$P_4'(t) = 0.02P_1(t) - 0.04P_4(t)$$

$$P_5'(t) = 0.01P_4(t) - 0.03P_5(t)$$

$$P_F'(t) = 0.05P_3(t) + 0.03P_5(t)$$

with the initial conditions are $P_1(0) = 1$ and others are 0, the system reliability function can be obtained as:

$$R(t) = 1 - P_F(t)$$

■

3.5. Notes and References

For reliability of hardware systems, Pukite & Pukite (1998) summarized some common configurations and implemented simple Markov models. Other than the Markov models, Elsayed (1996) described many other models that are commonly used in reliability engineering. There are also many general texts on reliability engineering and most of them deal with models for hardware systems.

Bobbio *et al.* (1980) first used Markov models in the study of a single hardware component that may contain multiple failure modes. Recently, Levitin *et al.* (1998) introduced a method called UGF (Universal Generating Function) in dealing with multiple failure modes. Alexopoulos & Shultes (2001) presented a method using an importance-sampling plan that dynamically adjusts the transition probabilities of the embedded Markov chain by attempting to cancel terms of the likelihood ratio within each cycle.

Kuo & Zuo (2003) summarized the reliability modeling for k-out-of-N configurations and presented the optimization schedules in improving the system reliability. Arulmozhi (2003) further presented a simple and efficient computational method for determining the reliability of k-out-of-N system whose components are of heterogeneous property. For parallel configurations, besides the majority voting and k-out-of-N voting introduced in this chapter, there are many other voting schemes, such as the enhanced voting scheme (Ammann & Knight, 1988), the weighted voting scheme (Levitin, 2001) and so on. Latif-Shabgahi *et al.* (2000) summarized various voting schemes for different fault tolerant systems. Chang *et al.* (2000) provided an extensive coverage on consecutive-k-out-of-n systems.

For the standby configurations, Sherwin & Bossche (1993) summarized the reliability analysis for both hot (active) standby and cold standby systems. Later, Chen *et al.* (1994) studied the reliability of a warm standby system which is an intermediate case between the hot and cold standby. Recently, Zhao & Liu (2003) provided a unified modeling idea for both parallel and standby redundancy optimization problems based on the system reliability analysis.

CHAPTER 4

MODELS FOR SOFTWARE RELIABILITY

Software is an important element in computing systems. Different from hardware, the software does not wear-out and it can be easily reproduced. Furthermore, software systems are usually debugged during the testing phase so that their reliability is improving over time as a result of detecting and removing software faults. Many software reliability growth models have been proposed for the study of software reliability, e.g. Xie (1991), Lyu (1996) and Pham (2000).

Markov models are one of the first types of models proposed in software reliability analysis. This chapter mainly summarizes models of this type. In addition, Nonhomogeneous Poisson Process (NHPP) models, which are important in software reliability analysis, are also discussed in this chapter.

4.1. Basic Markov Model

The basic Markov model in software reliability is the model originally developed by Jelinski & Moranda (1972). It is one of the earliest models and many later Markov models which can be considered as modifications or extensions of this basic Markov model.

4.1.1. Model description

The underlying assumptions of the Jelinski-Moranda (JM) model are:

1) The number of initial software faults is an unknown but fixed constant.
2) A detected fault is removed immediately and no new faults are introduced.
3) Times between failures are independent, exponentially distributed random variable.
4) All remaining faults in the software contribute the same amount to the software failure rate.

The initial number of faults in the software before the testing starts is denoted by N_0. From the assumptions (3) and (4), the initial failure rate is then equal to $N_0 \cdot \phi$, where ϕ is a constant of proportionality denoting the failure rate contributed by each fault. It follows from the assumption (2) that, after a new fault is detected and removed, the number of remaining faults is decreased by one. Hence after the *i*:th failure, there are $N_0 - i$ faults left, and the failure rate decreases to $\phi(N_0 - i)$. This Markov process is depicted by Fig. 4.1 where state *k* means that there are *k* faults left in the software.

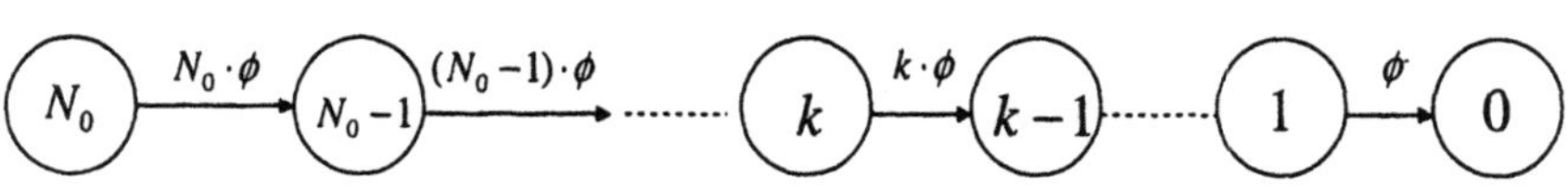

Fig. 4.1. Markov process of Jelinski-Moranda model.

The *i*:th failure-free period, i.e., the time between the (*i*-1):st and the *i*:th failure is denoted by T_i, $i = 1,2,\ldots,N_0$. By the assumptions, T_i's are then exponentially distributed random variables with parameter

$$\lambda(i) = \phi[N_0 - (i-1)] = \phi(N_0 - i + 1), \quad i = 1,2,\dots,N_0 \tag{4.1}$$

The distribution of T_i is given by

$$P(T_i < t_i) = \phi(N_0 - i + 1)\exp\{-\phi(N_0 - i + 1)t_i\}, \quad i = 1,2,\dots,N_0 \tag{4.2}$$

The main property of the JM-model is that the failure rate is constant between the detection of two consecutive failures. It is reasonable if the software is unchanged and the testing is random and homogeneous.

4.1.2. Parameter estimation

The parameters of the JM-model may easily be estimated by using the method of maximum likelihood. Let t_i denotes the observed *i*:th failure-free time interval during the testing phase. The number of faults detected is denoted here by n which will be called the sample size. If a failure time data set $\vec{t} = \{t_1, t_2, \dots, t_n; n > 0\}$ is given, the parameters ϕ and N_0 in the JM-model can be estimated by maximizing the likelihood function.

The likelihood function of the parameters ϕ and N_0 is given by

$$L(t_1, t_2, \dots; N_0, \phi) = \prod_{i=1}^{n} \phi(N_0 - i + 1)\exp\{-\phi(N_0 - i + 1)t_i\}$$

$$= \phi^n \prod_{i=1}^{n} (N_0 - i + 1)\exp\left\{-\phi\sum_{i=1}^{n}(N_0 - i + 1)t_i\right\} \tag{4.3}$$

The natural logarithm of the above likelihood function is

$$\ln L = \ln\left[\phi^n \prod_{i=1}^{n} (N_0 - i + 1)\exp\left\{-\phi\sum_{i=1}^{n}(N_0 - i + 1)t_i\right\}\right]$$

$$= n\ln\phi + \sum_{i=1}^{n}\ln(N_0 - i + 1) - \phi\sum_{i=1}^{n}(N_0 - i + 1)t_i \tag{4.4}$$

By taking the partial derivatives of this log-likelihood function above with respect to N_0 and ϕ, respectively, and equating them to zero, the following likelihood equations can be obtained,

$$\frac{\partial \ln L}{\partial N_0}=\sum_{i=1}^{n}\frac{1}{N_0-i+1}-\sum_{i=1}^{n}\phi\cdot t_i=0 \tag{4.5}$$

and

$$\frac{\partial \ln L}{\partial \phi}=\frac{n}{\phi}-\sum_{i=1}^{n}(N_0-i+1)t_i=0 \tag{4.6}$$

By solving ϕ from Eq. (4.6), we get

$$\phi=n\left[\sum_{i=1}^{n}(N_0-i+1)t_i\right]^{-1} \tag{4.7}$$

and by inserting this into Eq. (4.5), we obtain an equation independent of ϕ as

$$\frac{1}{N_0}+\frac{1}{N_0-1}+\ldots+\frac{1}{N_0-n+1}=\frac{n\sum_{i=1}^{n}t_i}{\sum_{i=1}^{n}(N_0-i+1)t_i} \tag{4.8}$$

An estimate of N_0 can then be obtained by solving this equation. Inserting the estimated value into Eq. (4.7), we obtain an maximum likelihood estimate (MLE) of ϕ.

Example 4.1. Suppose that a software product is being tested by a group. Each time a failure is observed, the fault causing the failure is removed. The 30 test data of time between failures are recorded in Table 4.1.

Substituting the data of Table 4.1 into the likelihood equations, and solving them, we obtain $N_0=54$ and $\phi=0.00077$.

After the 30th failure, the estimated number of remaining faults is $k = 54 - 30 = 21$ and the failure rate at that time is

$$\lambda = k \cdot \phi = 21 \times 0.00077 = 0.0162$$

Table 4.1. A set of failure data.

Failure number	Time between failures	Failure number	Time between failures	Failure number	Time between failures
1	14.75	11	7.99	21	64.25
2	43.99	12	28.09	22	40.90
3	9.89	13	11.80	23	3.07
4	0.07	14	1.78	24	0.75
5	5.70	15	12.50	25	13.36
6	7.89	16	73.08	26	23.02
7	28.79	17	42.60	27	143.31
8	170.15	18	9.18	28	55.46
9	26.83	19	49.43	29	75.57
10	36.15	20	9.19	30	34.31

The estimated reliability function after the 30th failures is

$$R(t) = \exp(-0.0162t)$$

and the MTTF after the 30th failures is estimated as

$$\text{MTTF} = \frac{1}{\lambda} = 61.84$$

■

Note that the estimation of the number of initial faults might be unreasonable. Usually more failure data should be accumulated for an estimate to be accurate.

4.2. Extended Markov Models

In many cases, the basic Markov model (JM-model) is not accurate enough. Several of the assumptions may not be realistic. For example, software faults are not of the same size in a sense that some affect more input data than others do, and some faults are easier to be detected than others. Many extended models, which relax some assumptions of the JM-model, are proposed (Xie, 1991). Some of them are discussed in this section.

4.2.1. Proportional models

Moranda (1979) presented an extended Markov model whose basic assumptions are same as JM-model except assuming that the (*i*+1):st failure rate is proportional to the *i*:th failure rate, i.e.

$$\lambda_{i+1} = C_i \lambda_i, \quad i=0,1,2,\ldots \tag{4.9}$$

This Markov process can be depicted by the Markov chain shown in Fig. 4.2, where state *i* represents that *i* failures have occurred.

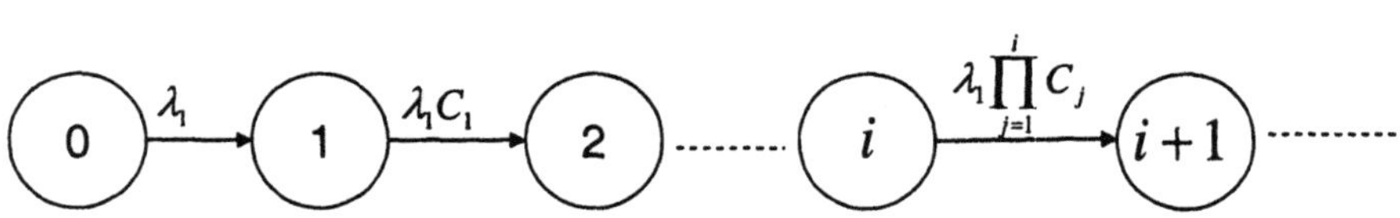

Fig. 4.2. The Markov chain for the proportional model.

This kind of model is called proportional model in Gaudoin *et al.* (1994). The idea is to consider that the difference between two successive failure rates is due only to the debugging, and practical constraints lead us to believe that the effect of this debugging is multiplicative. A proportional model is completely defined, given the distributions of λ_1 and $C=\{C_1, C_2, C_3, \ldots\}$.

Deterministic proportional model

In the simplest proportional model, all random variables are deterministic, i.e., λ_1 and C are constant. Hence it is called Deterministic Proportional Model as defined below.

Definition 4.1. The *Deterministic Proportional Model*, with parameters λ and θ, is the software reliability model where the random variable T_i are independent and exponentially distributed with parameter

$$\lambda \cdot \exp\{-(i-1)\theta\}, \quad i \geq 1 \tag{4.10}$$

This model was originally suggested by Moranda (1979) as geometric de-eutrophication model. Its detailed statistical property was studied by Gaudoin & Soler (1992) and we summarize some results here.

For the sake of convenience, let $C = \exp(-\theta)$, where θ is a real number. In fact, θ represents the quality of the debugging. If no debugging is done at all ($\theta = 0$), the failure rate remains constant; if the debugging is successful ($\theta > 0$), the failure rate decreases, and then the reliability grows, etc. The parameter λ is a scale parameter, and it is given by

$$\lambda = 1/E\{T_1\} \tag{4.11}$$

The likelihood for the observation of the first n times-between-failures t_i (i=1,2,...n) is:

$$L_{t_1,\ldots,t_n}(\lambda,\theta) = \prod_{i=1}^{n} \lambda_i \exp\{-\lambda_i t_i\}$$

$$= \lambda^n \exp\left\{-\left[\frac{n(n-1)}{2}\theta + \lambda \sum_{i=1}^{n} \exp\{-(i-1)\theta\} t_i\right]\right\} \tag{4.12}$$

Consequently, the maximum likelihood estimates of θ and λ, $\hat{\theta}(t_1,\cdots,t_n)$ and $\hat{\lambda}(t_1,\cdots,t_n)$ are the solution of the following equations

$$\frac{n}{\hat{\lambda}} = \sum_{i=1}^{n} t_i \exp\{-(i-1)\hat{\theta}\} \tag{4.13}$$

and

$$\sum_{i=1}^{n} (n-2i+1)t_i \exp\{-(i-1)\hat{\theta}\} = 0 \tag{4.14}$$

This equation expresses that $\hat{c} = \exp(-\hat{\theta})$ is a root of the polynomial of degree n-1, i.e.

$$\sum_{i=1}^{n} (n-2i+1)t_i c^{i-1} = 0 \tag{4.15}$$

Example 4.2. Suppose a software system is tested by a group. The 30 test data of time between failures are recorded in Table 4.2.

Table 4.2. 30 test data for time between failures.

Failure number	Time between failures	Failure number	Time between failures	Failure number	Time between failures
1	18.45	11	0.23	21	16.92
2	16.88	12	94.57	22	7.62
3	0.26	13	41.19	23	19.10
4	14.63	14	1.41	24	5.98
5	0.49	15	2.49	25	87.33
6	23.08	16	49.02	26	7.23
7	34.17	17	21.88	27	16.08
8	0.56	18	15.73	28	21.80
9	7.92	19	21.18	29	103.30
10	6.86	20	64.94	30	3.35

To estimate the parameters of λ and θ for the Deterministic Proportional Model, substitute the data of Table 4.2 into the above equations, and solve them to obtain $\hat{\theta} = 0.038$ and $\hat{\lambda} = 0.0754$.

The mean time to failure (MTTF) after the 30th failures, $E(T_{31})$, is

$$\text{MTTF} = E(T_{31}) = \frac{1}{0.0754\exp\{-(31-1)\cdot 0.038\}} = 41.47 \text{ (hours)}$$

If the customers require the MTTF of the software product should be no less than 70 hours, i.e., $E(T_i) \geq 70$, then

$$E(T_i) = \frac{1}{0.0754\exp\{-0.038(i-1)\}} \geq 70$$

Solving this, we get $i \geq 44.7$, so that the number of removed faults need to be at least 45. That is, at least $45 - 30 = 15$ more faults need to be removed.

The expected time for further detecting/removing the additional 11 faults is

$$\sum_{j=31}^{45} \frac{1}{0.0754\exp\{-0.038(j-1)\}} = 822.6 \text{ (hours)}$$

This is an estimated additional testing time needed.

■

Lognormal proportional model

In fact, the assumption of Deterministic Proportional Model that the C_i (mean quality) is constant, is not realistic. A more realistic assumption would be that the mean qualities of the successive debugging are independent random variables Q_i with a homogeneous normal distribution. Then,

$$C_i = \exp(-Q_i)$$

is a lognormal distribution. Gaudoin *et al.* (1994) presented a lognormal proportional model with

$$\lambda_{i+1} = \exp(-Q_i)\lambda_i \tag{4.16}$$

in which Q_i is normally distributed with mean θ and standard deviation σ.

The mean and variance of T_i are derived by Gaudoin *et al.* (1994) as:

$$E\{T_i\} = \frac{1}{\lambda}\exp\left\{(i-1)(\theta + \frac{\sigma^2}{2})\right\}$$

and

$$Var\{T_i\} = \frac{1}{\lambda^2}\exp\{2(i-1)(\theta + 0.5\sigma^2)\}\cdot[2\exp\{(i-1)\cdot\sigma^2\} - 1]$$

4.2.2. DFI (Decreasing Failure Intensity) model

A serious critique of the JM-model is that not all software faults contribute to the same amount of the failure rate. Some generalizations and modifications of the JM-model are presented in Xie (1987). We briefly describe this general formulation together with some special cases in this section.

General DFI formulation

The JM-model can be modified by using other function for $\lambda(i)$. Note that $\lambda(i)$ is defined as the rate of the occurrence of next failure after the removal of *i*-1 faults. The failure intensity is DFI (Decreasing Failure Intensity) if $\lambda(i)$ is a decreasing function of *i*. A DFI model is thus a Markov counting process model with decreasing failure intensity.

Under the general assumptions above, the cumulative number of faults detected and removed, $\{N(t), t \geq 0\}$, is a Markov process with decreasing failure rate $\lambda(i)$. The theory for CTMC can be applied.

If $P_i(t) = P\{N(t) = i\}$, $i = 0,1,\ldots,N_0$, the Chapman-Kolmogorov equations are given as

$$P_0'(t) = -\lambda(1)P_0(t)$$

$$P_i'(t) = -\lambda(i+1)P_i(t) + \lambda(i)P_{i-1}(t),\quad i = 2,3,\ldots,N_0 - 1 \tag{4.17}$$

$$P_{N_0}'(t) = -\lambda(N_0)P_{N_0-1}(t)$$

with the initial conditions

$$P_0(0) = 1 \text{ and } P_i(0) = 0 \text{ for } i > 0$$

The above equations can easily be solved and the solution is as follows (Xie, 1991),

$$P_0(t) = \exp\{-\lambda(1)t\}$$

$$P_1(t) = \frac{\lambda(1)}{\lambda(2) - \lambda(1)}(e_1 - e_0)$$

$$P_i(t) = \sum_{j=0}^{N_0-1} A_j^{(N_0-1)} e_j,\quad i = 2,3,\ldots,N_0 - 1$$

and for $i = N_0$, we have

$$P_{N_0}(t) = -\sum_{j=0}^{N_0-1} A_j^{(N_0-1)} \frac{\lambda(N_0)}{\lambda(j+1)} e_j$$

where the quantities e_j, j=0,1,..., $N_0 - 1$, are defined as

$$e_j = \exp\{-\lambda(j+1)\cdot t\}, \quad j=0,1,\ldots,\ N_0-1$$

and $A_j^{(i)}$ can be calculated recursively through

$$A_j^{(i)} = \frac{\lambda(i)}{\lambda(i+1)-\lambda(j+1)} A_j^{(i-1)}, \quad j<i$$

$$A_i^{(i)} = -\sum_{j=0}^{i-1} A_j^{(i)}$$

Some specific DFI models

A direct generalization of the JM-model is to use a power-type function for the failure intensity function, $\lambda(i)$. The power type DFI Markov model was studied by Xie & Bergman (1988) assuming the failure rate

$$\lambda(i) = \phi[N_0-(i-1)]^\alpha, i=1,2,\ldots,N_0 \tag{4.18}$$

It is reasonable to assume that $\lambda(i)$ is a convex function of i and α is likely to be greater than one, since in this case, the decrease of the failure rate is larger at the beginning.

Another special case of the DFI model is the exponential-type Markov model which assumes that the failure rate is an exponential function of the number of remaining faults. It is characterized by the failure rate function

$$\lambda(i) = \phi[\exp\{-\beta(N_0-i+1)\}-1], \quad i=1,2,\ldots,N_0 \tag{4.19}$$

For the exponential-type DFI model, the decrease of the failure intensity at the beginning is much faster than that at a later phase.

It is interesting to note that some of the proportional models can also be attributed to DFI model. If all the $C_i<1$ $(i=1,2,\ldots,N_0)$ in a proportional model, the failure rate $\lambda(i)$ is actually a decreasing function to the number of remaining faults, which follows the DFI definition.

4.2.3. Time-dependent transition probability models

Sometimes the failure rate function depends not only on the number of detected faults i but also on the time t_i whose Markov process is shown as Fig. 4.3.

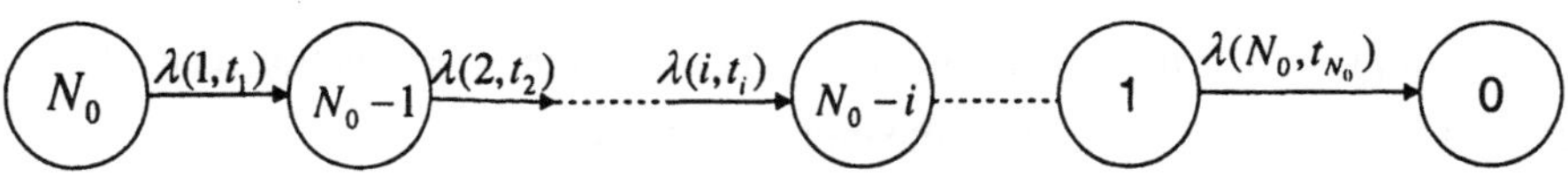

Fig. 4.3. CTMC for time dependent transition probability models.

There are several models which extend the JM-model by assuming that the probability of state change is also time-dependent. Schick-Wolverton model is one of the first models of this type (Schick & Wolverton, 1978). The general assumptions made by the Schick-Wolverton model are the same as those for the JM-model except that the times between failures are independent of the density function given by

$$f(t_i)=\phi(N_0-i+1)t_i\exp\left\{\frac{-\phi(N_0-i+1)t_i^2}{2}\right\},\quad i=1,2,\ldots,N_0 \tag{4.20}$$

in which N_0 is the number of initial faults and ϕ is another parameter.

Hence, the main difference between the Schick-Wolverton model and the JM-Model is that the times between failures are not exponential. In the Schick-Wolverton model the failure rate function to the detection of the i:th fault is

$$\lambda(i,t_i)=\phi(N_0-i+1)t_i \tag{4.21}$$

Note that the failure rate function of the Schick-Wolverton model depends both on i, the number of removed faults and on t_i, the time since the removal of last fault.

The Schick-Wolverton model with time-dependent failure rate was further extended by Shanthikumar (1981). Shanthikumar (1981) model supposes that there are N_0 initial software faults and assumed that after i faults are removed, the failure rate of the software is given by

$$\lambda(i,t) = \phi(t)(N_0 - i + 1)\, , \; i=1,2,\ldots \tag{4.22}$$

where $\phi(t)$ is a proportionality factor. The parameter estimation can also be carried out using the method of maximum likelihood.

The Markov formulation and solution procedures of Shanthikumar (1981) model are briefly introduced here. Denote by $P_i(t)$ the probability distribution function of $N(t)$, the number of faults that are detected and removed during time $[0,t)$. Under the Markovian assumption, we have that the forward Kolmogorov's differential equations are given as follows,

$$\frac{dP_0(t)}{dt} = -N_0\phi(t)P_0(t)$$

$$\frac{dP_i(t)}{dt} = (N_0 - i + 1)\phi(t)P_{i-1}(t) - (N_0 - i)\phi(t)P_i(t)\, ; \quad 1 \le i \le N_0$$

The initial conditions are

$$P_i(0) = 0\, , \;\; i > 0 \;\text{ and }\; P_0(0) = 1$$

The equations can easily be solved and the solution is given by

$$P_i(t) = \binom{N_0}{i}\left[\exp\left\{-\int_0^t \phi(x)dx\right\}\right]^{N_0 - i}\left[1 - \exp\left\{-\int_0^t \phi(x)dx\right\}\right]^i ; \;\; 0 \le i \le N_0 \tag{4.23}$$

4.2.4. Imperfect debugging models

The imperfect removal of a detected fault is a common situation in practice and the JM-model does not take this into account. This section extends the JM-model by relaxing the assumption of perfect debugging process. During an imperfect debugging process, there are two kinds of imperfect removal:

1) the fault is not removed successfully while no new fault is introduced.
2) the fault is not removed successfully while new faults are generated due to the incorrect diagnoses.

For the former type of imperfect removal, the process is still a monotonous death process in terms of the number of remaining faults; while the latter one is a birth-death process in terms of the number of remaining faults. Both types of imperfect debugging models will be discussed in the following.

Monotonous death process

Goel (1985) suggested a Markov model by assuming that each detected fault is removed with probability p. Hence, with probability q=1-p, a detected fault is not perfectly removed and the quantity q can be interpreted as the imperfect debugging probability. This process can be modeled by a DTMC as depicted by Fig. 4.4 where i is the number of detected failures.

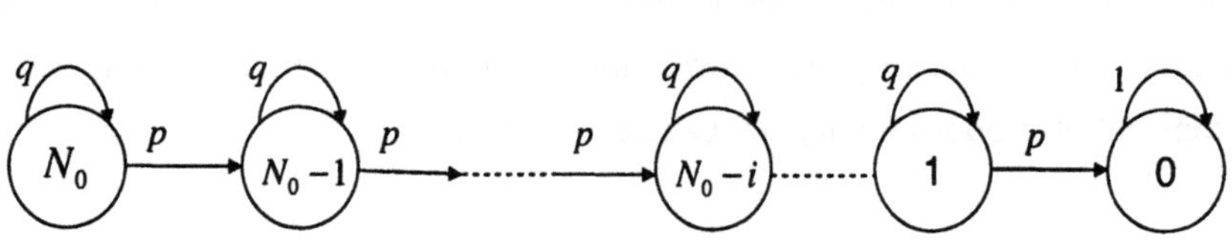

Fig. 4.4. DTMC for the monotonous death process of imperfect debugging model.

The counting process of the cumulative number of detected faults at time t is modeled as a Markov process with transition probability depending on the probability of imperfect debugging. Still it is assumed that times between the transitions are exponential with a parameter which depends only on the number of remaining faults. After the occurrence of i-1 failures, $p\cdot(i-1)$ faults are removed on the average. Hence, approximately, there are $N_0 - p(i-1)$ faults left, where N_0 denotes the number of initial faults as before. The failure rate between the (i-1):st and the i:th failures is then

$$\lambda(i) = \phi[N_0 - p(i-1)] \tag{4.24}$$

Using this transition function, other reliability measures can be calculated as for the JM-model. Note that the above rate function can be rewritten as

$$\lambda(i) = \phi\cdot p\left[\frac{N_0}{p} - (i-1)\right] \tag{4.25}$$

and from this it can be seen that it is just the same as that for the JM-model with ϕ replaced by $\phi\cdot p$ and N_0 replaced by N_0/p.

As a consequence, p, N_0 and ϕ are indistinguishable. However, $\phi\cdot p$ and N_0/p can still be estimated similar to that for the parameters in the JM-model and N_0/p can be interpreted as the expected number of failures that will eventually occur. Another advantage of using this model is when we know the probability of imperfect debugging, p. For example, from the previous experience or by checking after correction, the number of initial faults N_0 and the constant of proportionality ϕ can be estimated.

Example 4.3. Suppose that a software product is being tested by a group. The 30 test data of time between failures are recorded in Table 4.3.

If the software failures follow the above imperfect debugging model given p=0.9, viewing it as the JM-model first, we get the following estimates

$$N_0 / p = 42.56 \quad \text{and} \quad \phi \cdot p = 0.00104$$

Substituting p=0.9, we get

$$N_0 = 38 \quad \text{and} \quad \phi = 0.00116$$

Table 4.3. 30 test data for time between failures.

Failure number	Time between failures	Failure number	Time between failures	Failure number	Time between failures
1	8.12	11	0.01	21	85.56
2	9.76	12	20.48	22	17.95
3	10.52	13	5.28	23	57.01
4	59.98	14	65.28	24	80.08
5	8.67	15	11.83	25	48.40
6	29.96	16	10.60	26	25.01
7	24.26	17	62.42	27	97.98
8	30.74	18	18.97	28	58.61
9	51.00	19	162.48	29	55.75
10	18.23	20	4.88	30	4.58

■

Birth-death process

Furthermore, if we allow the imperfect debugging process to introduce new faults into the software due to the wrong diagnoses or incorrect modifications, the debugging process becomes a birth-death Markov process. Kremer (1983) assumes that when a failure occurs, the fault content is assumed to be reduced by 1 with probability p, the fault content is not changed with probability q, and a new fault is generated with probability r. The obvious equality is that

$$p + q + r = 1$$

This implies that we have a birth-death process with a birth rate $v(t) = r \cdot \lambda(t)$ and a death rate $\mu(t) = p \cdot \lambda(t)$. It can be depicted by the CTMC as Fig. 4.5.

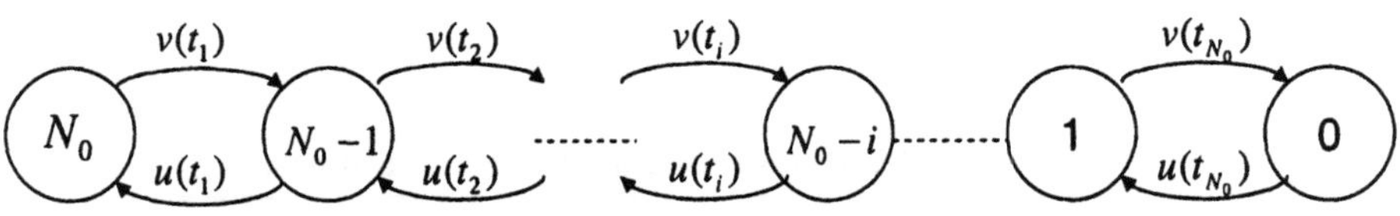

Fig. 4.5. CTMC for the birth-death process of imperfect debugging model.

However, in order to fit failure data and obtain further applicable results, assumptions on the failure rate function $\lambda(t)$ must be made.

Denoted by $N(t)$ the number of remaining faults in the software at time t and let

$$P_i(t) = \Pr\{N(t) = i\}, \quad i=0,1,\ldots,N_0.$$

We obtain the forward Kolmogorov equations of this Markov process as

$$P'_i(t) = (i-1)v(t)P_{i-1}(t) - i[v(t)+\mu(t)]P_i(t) + (i+1)\mu(t)P_{i+1}(t), \quad i \geq 0 \tag{4.26}$$

Generally, by inserting $v(t)$ and $\mu(t)$ and using the initial condition $P_{N_0}(0) = 1$, the differential equations can be solved by using the probability generating function suggested in Kremer (1983).

Imperfect debugging model considering multi-type failure

In practice, software failures can be classified into different types according to their severity or characteristics. Different types of failures may cause different software reliability performance. Tokuno & Yamada (2001) presented a Markov model with two types of failures that have different kinds of failure rates and imperfect debugging process. The first type is the failures caused by faults originally latent in the system prior to the testing, denoted by F1. The second type is the failures due to faults randomly introduced or regenerated during the testing phase, denoted by F2.

They assumed that

1) The failure rate for Fl is constant between failures and decreases geometrically as each fault is corrected, and the failure rate for F2 is constant throughout the testing phase.
2) The debugging activity for the fault is imperfect: denoted by p the probability for a fault to be removed successfully.
3) The debugging activity is performed without distinguishing between Fl and F2.
4) The probability that two or more software failures occur simultaneously is negligible.
5) At most one fault is corrected when the debugging activity is performed, and the fault-correction time is negligible or not considered.

Let $X(t)$ be a counting process representing the cumulative number of faults corrected up to testing time t. From the assumption 2, when i faults have been corrected by an arbitrary testing time t, after the next software failure occurs,

$$X(t) = \begin{cases} i, & \text{with probability } q \\ i+1, & \text{with probability } p \end{cases} \tag{4.27}$$

from the assumptions 1 and 3, when i faults have been corrected, the failure rate for the next software failure-occurrence is given by

$$\lambda(i) = D \cdot k^i + \theta, \quad i = 0,1,2,\ldots, \quad D > 0, \quad 0 < k < 1, \quad \theta \geq 0 \tag{4.28}$$

where D is the initial failure rate for F1, k is the decreasing ratio of the failure rate, and θ is the failure rate for F2.

The reliability function to the next software failure is given by

$$R_i(t) = \exp\{-(D \cdot k^i + \theta)t\} \tag{4.29}$$

Furthermore, let $Q_{ij}(\tau)$ denote the one step transition probability that after making a transition into state *i*, the process $\{X(t), t \geq 0\}$ makes a transition into state *j* by time τ. Then, we have

$$Q_{ij}(\tau) = P_{ij} \cdot [1 - \exp\{-(D \cdot k^i + \theta)\tau\}] \tag{4.30}$$

where P_{ij} are the transition probabilities from state *i* to state *j*.

4.3. Modular Software Systems

If possible, the architecture of software should be taken into account instead of considering the software as a black-box system. Markov models can also be applied in analyzing the reliability for modular software system.

4.3.1. The Littlewood semi-Markov model

Littlewood (1979) incorporated the structure of the software into the Markov process using a kind of semi-Markov model. The program is assumed to be comprised of a finite number of modules and the transfer of control between modules is described by the probability

$$p_{ij} = \Pr\{\text{program transits from module } i \text{ to module } j\}$$

The time spent in each module has a general distribution $F_{ij}(t)$ which depends upon *i* and *j*, with finite mean m_{ij}. When module *i* is executed, failures occur according to a Poisson process with parameter λ_i. The transfer of control between modules has a probability v_{ij} of a failure.

The interest of the composite model is focused on the total number of failures of integrated software system in time interval $(0, t]$, denoted by *N*(*t*). The asymptotic Poisson process approximation for $N(t)$ is obtained under the assumption that failures are very infrequent. The times between failures tend to be much larger than the times between exchanges of control. The failure occurrence rate of this Poisson process is given by

$$\lambda_s = \sum_i a_i \lambda_i + \sum_{i,j} b_{ij} v_{ij} \tag{4.31}$$

where

$$a_i = \frac{\pi_i \sum_j p_{ij} m_{ij}}{\sum_i \pi_i \sum_j p_{ij} m_{ij}} \tag{4.32}$$

represents the proportion of time spent in module i, and

$$b_{ij} = \frac{\pi_i p_{ij}}{\sum_i \pi_i \sum_j p_{ij} m_{ij}} \tag{4.33}$$

is the frequency of transfer of control between i and j.

4.3.2. Some other modular software models

User-oriented model

Similar to the Littlewood semi-Markov model, a model called the *user-oriented model*, was developed by Cheung (1980) where the user profile can be incorporated into the modeling. The model is a Markov model based on the reliability of each individual module and the inter-modular transition probabilities as the user profile.

Assume that the program flow graph of a terminating application has a single entry and a single exit node, and that the transfer of control among modules can be described by an absorbing DTMC with a transition probability matrix $\boldsymbol{P}=\{p_{ij}\}$. Modules fail independently and the reliability of the module i is the probability R_i that the module performs its function correctly.

Two absorbing states $\boldsymbol{C}$ and $\boldsymbol{F}$ are added, representing the correct output and failure state, respectively, and the transition probability matrix $\boldsymbol{P}$ is modified appropriately to $\hat{P}$. The original transition probability p_{ij} between the modules i and j is modified to $R_i p_{ij}$. This represents the probability that the module i produces the correct result and the control is transferred to module j. From the exit state n, a directed edge to state C is created with transition probability R to represent the correct execution. The failure of a module i is considered by creating a directed edge to failure state F with transition probability $1-R_i$. Hence, DTMC defined with transition probability matrix $\hat{P}$ is a composite model of the software system. The reliability of the program is the probability of reaching the absorbing state C of the DTMC.

Let Q be the matrix obtained from $\hat{P}$ by deleting rows and columns corresponding to the absorbing states C and F. $Q^k(1,n)$ represents the probability of reaching state n from 1 through k transitions. From initial state 1 to final state n, the number of transitions k may vary from 0 to infinity. It can be show that

$$S = I + Q + Q^2 + Q^3 + \cdots = \sum_{k=0}^{\infty} Q^k = (I - Q)^{-1} \tag{4.34}$$

and it follows that the overall system reliability can be computed as

$$R = S(1,n) \cdot R_n \tag{4.35}$$

Task-oriented model

A modular software is usually developed to complete certain tasks. Kubat (1989) presented a *task-oriented model* which considered the case of a terminating software application composed of n modules designed for K different tasks. Each task may require several modules and the same module can be used for different tasks. Transitions between modules follow a DTMC such that with probability $q_i(k)$ task k will first call module i and with probability $p_{ij}(k)$ task k will call module j after executing in module i. The sojourn time during the visit in module i by task k has the density function $g_i(k,t)$. Hence, a semi-Markov process can be used.

The failure rate of module i is λ_i. As shown in Kubat (1989), the probability that no failure occurs during the execution of task k, while in module i is

$$R_i(k) = \int_0^{\infty} e^{-\lambda_i t} g_i(k,t)dt \tag{4.36}$$

The expected number of visits in module i by task k, denoted by $V_i(k)$, can be obtained by solving

$$V_i(k) = q_i(k) + \sum_{j=1}^{n} V_j(k) p_{ij}(k); \quad i = 1,2,...,n, \quad k = 1,2,...,K \tag{4.37}$$

The probability that there will be no failure when running for task k can be approximated by

$$R(k) \approx \prod_{i=1}^{n} [R_i(k)]^{V_i(k)} \tag{4.38}$$

and the system failure rate is calculated by

$$\lambda_s = \sum_{k=1}^{K} r_k [1 - R(k)] \tag{4.39}$$

where r_k is the arrival rate of task k.

Multi-type failure model in modular software

Ledoux (1999) further proposed a Markov models to include the multi-type failures into the modular software reliability analysis. They constructed an irreducible CTMC with transition rates q_{ij} to model the software composed of a set of components C. In the model, two types of failures are considered: primary failures and secondary failures. The primary failure leads to an execution break; the execution is restarted after some delay. A secondary failure does not affect the software because the execution is assumed to be restarted instantaneously when the failure appears. For an active component c_i, a primary failure occurs with constant rate λ_{i1}, while the secondary failures are described as Poisson process with rate λ_{i2}. When control is transferred between two components *i* and *j* then a primary (secondary) interface failure occurs with probability v_{ij1} (v_{ij2}).

Following the occurrence of a primary failure, a recovery state is occupied, and the delay of the execution break is a random variable with a phase type distribution. Denoting by *R* the set of recovery states, the state space becomes $C \cup R$. Hence, the CTMC that defines the architecture is replaced by a CTMC that models alternation of operational-recovery periods. The associated generator matrix defines the following transition rates: from c_i to c_j with no failure; from c_i to c_j with a secondary failure; from c_i to c_j with a primary failure; from recovery state *i* to recovery state *j*; and from recovery state *i* to c_j.

A Markov model is then constructed according to the architecture of different modules and their states. Based on the CTMC, the Chapman-Kolmogorov equations can be obtained and solved by computational tools.

4.4. Models for Correlated Failures

Perhaps the most stringent restriction in most software reliability models is the assumption of statistical independence among successive software failures. It is common for software failures to be correlated in successive runs. In order to deal with this issue, Goseva-Popstojanova & Trivedi (2000) formulated a Markov renewal model that can consider the phenomena of failure correlation.

4.4.1. Description of the correlated failures

Since each software run has two possible outcomes (success or failure), the usual way of looking at the sequence of software runs is to consider it as a sequence of Bernoulli trials, where each trial has success-probability p and failure-probability 1-p. Goseva-Popstojanova & Trivedi (2000) constructed a Markov renewal model for the sequence of dependent software runs in two stages:

1) Define a DTMC which considers the outcomes from the sequence of possibly dependent software runs in discrete time.
2) Construct the process in continuous time by attaching the distributions of the runs execution to the transitions of the DTMC.

The assumptions of the model are:

1) The probability of success or failure at each run depends on the outcome of the previous run.
2) A sequence of software runs is defined as a sequence of dependent Bernoulli trials.
3) Each software run takes a random amount of time to be executed.
4) Software execution times are not identically distributed for successful and failed runs.

4.4.2. Constructing the semi-Markov model

Associated with the j:th software-run, let Z_j be a random variable that distinguishes whether the outcome of that particular run resulted in success or failure:

$$Z_j = \begin{cases} 0 & \text{a success on run } j \\ 1 & \text{a failure on run } j \end{cases} \tag{4.40}$$

Here we use score 1 for each time a failure occurs and 0 otherwise. The number of runs that have resulted in a failure among n successive software runs is: $S_n = \sum_{j=1}^{n} Z_j$ of n possibly dependent random variables.

Suppose that if run j results in failure. At run (j+1), the failure probability is q and the success probability is $\overline{q}$. Similarly, if run j results in success, then p and $\overline{p}$ are the probabilities of success and failure, respectively, at run (j+1). The sequence of dependent Bernoulli trials $\{Z_j; j \geq 1\}$ defines a DTMC with 2 states. One is a success state denoted by 0; the other denoted by 1 is a failure. Its transition probability matrix is

$$P = \begin{bmatrix} p & \overline{p} \\ \overline{q} & q \end{bmatrix}$$

as shown by Fig. 4.6.

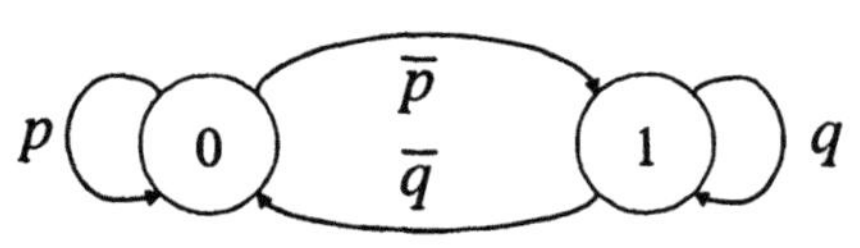

Fig. 4.6. DTMC interpretation of dependent Bernoulli trials.

The unconditional probability of failure on run (j+1) can be derived, see e.g. Goseva-Popstojanova & Trivedi (2000), as

$$\Pr\{Z_{j+1} = 1\} = \overline{p} + (p + q - 1) \cdot \Pr\{Z_j = 1\} \tag{4.41}$$

This equation shows the property of failure correlation in successive runs. If $p + q = 1$, the Markov chain describes a sequence of independent Bernoulli trials, and the above equation reduces to:

$$\Pr\{Z_{j+1} = 1\} = \overline{p} = q$$

which means that the failure probability does not depend on the outcome of the previous run. When $p+q>1$, runs are positively correlated, i.e. if a software failure occurs in run *j*, then there is an increased chance that another failure occurs in the next run. In this case, failures occur in clusters. Finally, when $p+q<1$, runs are negatively correlated. In this case, if a software failure occurs in run *j*, then there is an increased chance that a success occurs in run (*j*+1), i.e., there is a lack of clustering.

The next step in the model construction is to obtain a process in continuous time. Let $F_{kl}(t)$ be the cumulative distribution function of the time spent in a transition from state *k* to state *l*. It is realistic to assume that the runs execution times are not identically distributed for successful and failed runs. Hence, the $F_{kl}(t)$ depend only of the type of point at the end of the interval, i.e.,

$$F_{00}(t)=F_{10}(t)=F_{exS}(t)$$

and

$$F_{01}(t)=F_{11}(t)=F_{exF}(t)$$

With the addition of the $F_{kl}(t)$ to the transitions of DTMC we obtain a Markov renewal process as the software reliability model in continuous time.

4.4.3. Considering software reliability growth

During the testing phase, software is subjected to a sequence of runs, making no changes if there is no failure. When a failure occurs on any run, then an attempt is made to fix the underlying fault which causes the conditional probabilities of success and failure on the next run to change. The software reliability growth model in discrete time can be described with a sequence of dependent Bernoulli trials with step-dependent probabilities. The underlying stochastic process is a nonhomogeneous DTMC.

The sequence $S_n = \sum_{j=1}^{n} Z_j$ provides an alternate description of the software reliability growth model considered here. That is, $\{S_n\}$ defines the DTMC presented in Fig. 4.7.

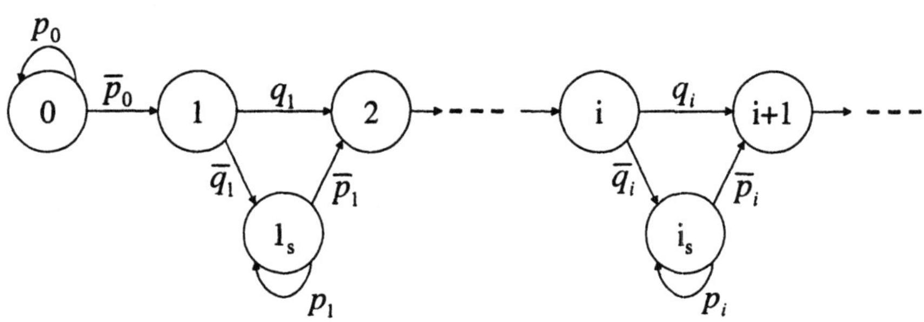

Fig. 4.7. States transition over time.

Both states i and i_s represent that the failure state has been occupied i times. State i represents the first trial for which $S_n = i$. State i_s represents all subsequent trials for which $S_n = i$, i.e., all subsequent successful runs before the occurrence of next failure (i+1). Without loss of generality let the first run be successful which means that 0 is the initial state.

The software reliability growth model in continuous time can also be obtained by assigning runs execution-time distributions to transitions of the DTMC in Fig. 4.7. For simplicity, we have chosen the same execution time distribution regardless of the outcome:

$$F_{00}(t) = F_{01}(t) = F_{10}(t) = F_{11}(t) = F(t)$$

Hence, T_{ex} of each software run has cumulative distribution function

$$F(t) = \Pr\{T_{ex} \le t\}$$

Let X_{i+1} be the number of runs between failures i and i+1, From Fig. 4.7, the random variable X_{i+1} $(i \geq 1)$ has the following distribution:

$$\Pr\{X_{i+1} = k\} = \begin{cases} q_i & k = 1 \\ \overline{q}_i \cdot p_i^{k-2} \overline{p}_i & k \geq 2 \end{cases} \tag{4.42}$$

It follows that the distribution function of the time to failure $(i+1)$, given that the system has had i failures, $i \geq 1$, is:

$$F_{i+1}(t) = q_i F(t) + \sum_{k=2}^{\infty} \overline{q}_i \cdot p_i^{k-2} \overline{p}_i F^{k*}(t) \tag{4.43}$$

where $F^{k*}(t)$ is k-fold convolution of $F(t)$.

The Laplace-Stieltjes transform of $F(t)$ is $\widetilde{F}(s)$, and then the above equation is transformed as

$$\begin{aligned} \widetilde{F}_{i+1}(s) &= q_i \widetilde{F}(s) + \sum_{k=2}^{\infty} \overline{q}_i \cdot p_i^{k-2} \overline{p}_i \widetilde{F}^k(s) \\ &= \frac{q_i \widetilde{F}(s) + (1 - p_i - q_i) \widetilde{F}^2(s)}{1 - p_i \widetilde{F}(s)} \end{aligned} \tag{4.44}$$

Its inversion is straightforward and reasonably simple closed-form results can be obtained when $F(t)$ has a rational Laplace-Stieltjes transform.

Some general properties of the inter-failure time can be developed without making assumptions about the form of $F(t)$. For example, the MTTF is:

$$MTTF = E[T_{i+1}] = -\frac{d\widetilde{F}_{i+1}(s)}{ds}\Big|_{s=0} = \frac{\overline{p}_i + \overline{q}_i}{\overline{p}_i} E[T_{ex}] \tag{4.45}$$

where $E[T_{ex}]$ is the mean execution-time.

Example 4.4. Suppose that the failures of a software are correlated between successive runs with p_0=0.7 (from success to success state) and q_0=0.8 (from failure to failure state). The execution time of each run is assumed to follow the exponential distribution with mean $\mu = 30$ hours.

Substituting the above values into the above equation, we get

$$MTTF = \frac{\bar{p}_0 + \bar{q}_0}{\bar{p}_0}\mu = \frac{0.3+0.2}{0.3}30 = 50 \quad \text{(hours)}$$

During the testing phase, when detecting a failure, we try to remove it, so the dependent probabilities are changing as Fig. 4.7. If we assume $p_{i+1} = 1.02p_i$ and $q_{i+1} = 0.98q_i$ then

$$p_i = (1.02)^i p_0 \text{ and } q_i = 0.98^i q_0, \quad i=1,2,\ldots$$

If the customer requires that the MTTF should be longer than 100 hours, to determine the testing time, we should use the following in equation

$$\frac{2-(1.02)^i 0.7 - (0.98)^i 0.8}{1-(1.02)^i 0.7} \cdot 30 \geq 100$$

Solving this, we get $i \geq 9.8$, so the least number of detected/debugged failures should be 10. Then, the expected testing time before release can be computed as

$$T \geq 30 \cdot \sum_{i=0}^{9} \frac{2-(1.02)^i 0.7 - (0.98)^i 0.8}{1-(1.02)^i 0.7} = 668.55 \text{ (hours)}$$

That is, in order to satisfy the customer requirement, the software should be tested for at least 669 hours before release.

■

4.5. Software NHPP Models

Although some basic and advanced Markov models are presented in the previous sections, some NHPP models are mentioned here due to their significant impact on the software reliability analysis. Such a model simply models the failure occurrence rate as a function of time (see e.g., Section 2.4). Hopefully this occurrence rate is decreasing when faults are removed as an effect of debussing. Note that after the release, the failure occurrence rate should be a constant unless the debugging is continued (Yang & Xie, 2000).

4.5.1. The Goel-Okumoto (GO) model

In 1979, Goel and Okumoto presented a simple model for the description of software failure process by assuming that the cumulative failure process is NHPP with a simple mean value function. Although NHPP models have been studied before, see e.g. Schneidewind (1975), the GO-model is the basic NHPP model that later has had a strong influence on the software reliability modeling history.

Model description

The general assumptions of the GO-model are

1) The cumulative number of faults detected at time t follows a Poisson distribution.
2) All faults are independent and have the same chance of being detected.
3) All detected faults are removed immediately and no new faults are introduced.

Specifically, the GO-model assumes that the failure process is modeled by an NHPP model with mean value function $m(t)$ given by

$$m(t) = a[1-\exp(-bt)], \quad a>0, b>0 \tag{4.46}$$

The failure intensity function can be derived by

$$\lambda(t) = \frac{d}{dt} m(t) = ab\exp(-bt) \tag{4.47}$$

where a and b are positive constant. Note that $m(\infty) = a$. The physical meaning of parameter a can be explained as the expected number of faults which are eventually detected. The quantity b can be interpreted as the failure occurrence rate per fault.

The expected number of remaining faults at time t can be calculated as

$$E[N(\infty) - N(t)] = m(\infty) - m(t) = a\exp(-bt)$$

The GO-model has a simple but interesting interpretation based on a model for fault detection process. Suppose that the expected number of faults detected in a time interval $[t, t + \Delta t)$ is proportional to the number of remaining faults, we have that

$$m(t + \Delta t) = b[a - m(t)]\Delta t$$

where b is a constant of proportionality.

The above difference equation can be transformed into a differential equation. Divide both sides by Δt and take limits by letting Δt tend to zero, we get the following equation,

$$m'(t) = a \cdot b - b \cdot m(t)$$

It can be shown that the solution of this differential equation, together with the initial condition $m(0) = 0$, lead to the mean value function of the GO-model.

Note that both the GO-model and JM-model give the exponentially decreasing number of remaining faults. It can be shown that these two models cannot be distinguished using only one realization from each model. However, the models are different because the JM-model assumes a discrete change of the failure intensity at the time of the removal of a fault while the GO-model assumes a continuous failure intensity function over the whole time domain.

Parameter estimation

Denoted by n_i the number of faults detected in time interval $[t_{i-1}, t_i)$, where $0 = t_0 < t_1 < \cdots < t_k$ and t_i are the running times since the software testing begins. The estimation of model parameters a and b can be carried out by maximizing the likelihood function, see e.g. Goel & Okumoto (1979). The likelihood function can be reduced to

$$\sum_{i=1}^{k} \frac{n_i[t_i \exp(-bt_i) - t_{i-1} \exp(-bt_{i-1})]}{\exp(-bt_{i-1}) - \exp(-bt_i)} = \frac{t_k \exp(-bt_k) \cdot \sum_{i=1}^{k} n_i}{1 - \exp(-bt_k)} \tag{4.48}$$

Solving this equation to calculate the estimate of b, and then a can be estimated as

$$a = \frac{\sum_{i=1}^{k} n_i}{1 - \exp(-bt_k)} \tag{4.49}$$

Usually, the above two equations has to be solved numerically. It can also be shown that the estimates are asymptotically normal and a confidence region can easily be established. A numerical example is illustrated below.

Example 4.5. Suppose a software product is being tested by a group. Each time when detecting the failure, it is removed and the time for repair is not computed in the test time. The 30 test data of time to failures are recorded in Table 4.4.

Solving the likelihood equations, we get $b = 0.0008$ and $a = 57$. The failure intensity function and the mean value function for this GO model are

$$\lambda(t) = 0.0456 \exp(-0.0008t)$$

and

$$m(t) = 57[1 - \exp(-0.0008t)]$$

Table 4.4. A set of time to failure data.

Failure number	Time to failures	Failure number	Time to failures	Failure number	Time to failures
1	1.33	11	288.43	21	547.21
2	3.43	12	288.84	22	554.65
3	24.87	13	303.02	23	629.93
4	58.15	14	330.30	24	741.44
5	85.78	15	375.16	25	773.25
6	145.84	16	414.85	26	789.56
7	203.84	17	417.96	27	815.74
8	205.82	18	434.56	28	874.62
9	219.97	19	517.28	29	888.81
10	244.09	20	543.72	30	924.94

■

4.5.2. S-shaped NHPP models

The mean value function of the GO-model is exponential-shaped. Based on the experience, it is observed that the curve of the cumulative number of faults is often S-shaped as shown by Fig. 4.8, see e.g. Yamada *et al.* (1984).

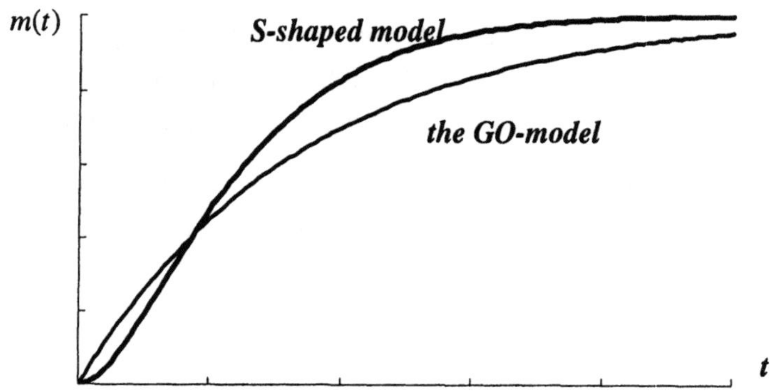

Fig. 4.8. The S-shaped mean value function.

This can be explained by the fact that at the beginning of the testing, some faults might be "covered" by other faults. Removing a detected fault at the beginning does not reduce the failure intensity very much since the same test data will still lead to a failure caused by other faults. Another reason of the S-shaped behavior is the learning effect as indicated in Yamada *et al.* (1984).

Delayed S-shaped NHPP model

The mean value function of the *delayed S-shaped* NHPP model is

$$m(t) = a[1-(1+bt)\exp(-bt)]\,;\ \ b>0\,, \tag{4.50}$$

This is a two-parameter S-shaped curve with parameter a denoting the number of faults to be detected and b corresponding to a fault detection rate. The corresponding failure intensity function of this delayed S-shaped NHPP model is

$$\lambda(t) = \frac{dm(t)}{dt} = ab(1+bt)\exp(-bt) - ab\exp(-bt) = ab^2 t\exp(-bt)$$

The expected number of remaining faults at time t is then

$$m(\infty) - m(t) = a(1+bt)\exp(-bt)$$

Inflected S-shaped NHPP model

The mean value function of the *inflected S-shaped* NHPP model is

$$m(t) = \frac{a[1-\exp(-bt)]}{1+c\exp(-bt)}\,;\ \ b>0, c>0$$

In the above a is again the total number of faults to be detected while b and c are called the fault detection rate and the inflection factor, respectively. The intensity function of this inflected S-shaped NHPP model can easily be derived as

$$\lambda(t) = \frac{dm(t)}{dt} = \frac{ab(1+c)\cdot\exp(-bt)}{[1+c\exp(-bt)]^2}$$

Given a set of failure data, for both delayed and inflated S-shaped NHPP models, numerical methods have to be used to solve the likelihood equation so that estimates of the parameters can be obtained.

4.5.3. Some other NHPP models

Besides the S-shaped models, there are many other NHPP models that extend the GO-model for different specific conditions.

Duane model

The Duane model assumes that the mean value function satisfies

$$m(t) = \left(\frac{t}{\alpha}\right)^{\beta}, \quad \alpha > 0, \ \beta > 0 \tag{4.51}$$

In the above, α and β are parameters which can be estimated by using collected failure data. The mean value functions with $\alpha = 100$ and different $\beta = \{0.5,1,2\}$ are depicted by the Fig. 4.9.

It can be noted that when $\beta = 1$, the Duane NHPP model is reduced to a Poisson process whose mean value function is a straight line. In such a case, there is no reliability growth. In fact, the Duane model can be used to model both reliability growth ($\beta < 1$) and reliability deterioration ($\beta > 1$) which is common in hardware systems.

The failure intensity function, $\lambda(t)$, is

$$\lambda(t) = \frac{d}{dt}m(t) = \frac{\beta}{\alpha}\left(\frac{t}{\alpha}\right)^{\beta-1}, \quad \alpha > 0, \ \beta > 0$$

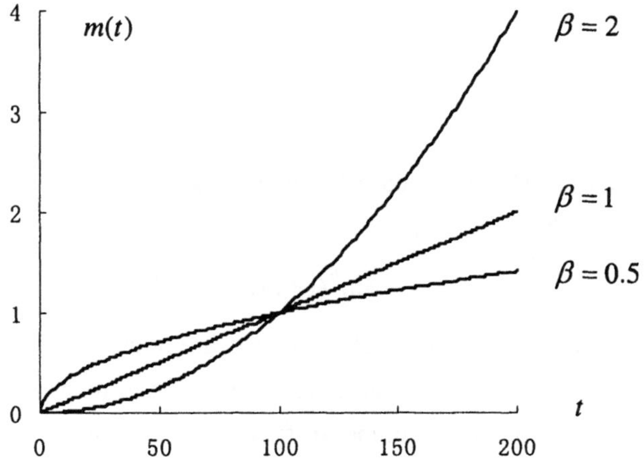

Fig. 4.9. Mean value functions of Duane NHPP models

One of the most important advantages of the Duane model is that if we plot the cumulative number of failure versus the cumulative testing time on a log-log-scale, the plotted points tends to be close to a straight line if the model is valid. This can be seen from the fact that the relation between $m(t)$ and t can be rewritten as

$$\ln m(t) = -\beta \ln \alpha + \beta \ln t = a + b \ln t$$

where $a = -\beta \ln \alpha$ and $b = \beta$. Hence, $\ln m(t)$ is a linear function of $\ln t$ and due to this linear relation, the parameters α and β may be estimated graphically and the model validity can easily be verified. In fact, this is called first-model-validation-then-parameter-estimation approach (Xie & Zhao, 1993).

The Duane model gives an infinite failure intensity at time zero. Littlewood (1984) proposed a *modified Duane model* with the mean value function

$$m(t) = k\left[1 - \left(\frac{\alpha}{\alpha + t}\right)^{\beta}\right], \quad \alpha > 0, \ \beta > 0, \ k > 0$$

The parameter k can be interpreted as the number of faults eventually to be detected.

Log-power model

Xie & Zhao (1993) presented a log-power model. The mean value function of this model can be written as

$$m(t) = a\ln^b(1+t)\,;\quad a,b>0\,,\quad t\ge 0 \tag{4.52}$$

This model has shown to be useful for software reliability analysis as it is a pure reliability growth model. It is also easy to use due to its graphical interpretation. The plot of the cumulative number of failures at time t against t+1 will tend to be a straight line on a log-double-log scale if the failures follow the log-power model. This can be seen from the following relationship

$$\ln m(t) = \ln a + b\ln\ln(1+t)$$

The slope of the fitted line gives an estimation of b and its intercept on the vertical axis gives an estimation of $\ln a$.

The failure intensity function of the log-power model can be obtained as

$$\lambda(t) = \frac{ab\ln^{b-1}(1+t)}{1+t}\,,\quad t\ge 0 \tag{4.53}$$

The failure intensity function is interesting from a practical point of view. The log-power model is able to analyze both the case of strictly decreasing failure intensity and the case of increasing-then-decreasing failure intensity function. For example, if $b\le 1$, then $\lambda(t)$ of the above equation is a monotonic decreasing function of t; Otherwise given $b>1$, $\lambda(t)$ is increasing if $0\le t<\exp(b-1)$ and decreasing if $t\ge \exp(b-1)$.

The estimation of the parameters a and b is also simple. Suppose total n failures are detected during the a testing period $(0,T]$ and the times to failures are ordered by $0 < t_1 < t_2 < \cdots < t_n \leq T$. The maximum likelihood estimation of a and b is then given by:

$$\hat{b} = \frac{n}{n \ln \ln(1+T) - \sum_{i=1}^{n} \ln \ln(1+t_i)}$$

and

$$\hat{a} = \frac{n}{\ln^{\hat{b}}(1+T)}$$

They can be simply calculated without numerical procedures.

Musa-Okumoto model

Musa and Okumoto (1984) is another model for infinite failures. This NHPP model is also called the logarithmic Poisson model. The mean value function is

$$m(t) = a \ln(1+bt),\ t > 0 \tag{4.54}$$

The failure intensity function is derived as

$$\lambda(t) = \frac{ab}{1+bt}$$

Given a set of failure time data $\{t_i, i = 1,2,...,n\}$, the maximum likelihood estimates of the parameters are the solutions of the following equations:

$$\begin{cases} \hat{a} = \dfrac{n}{\ln(1+\hat{b}t_n)} \\ \dfrac{1}{\hat{b}} \displaystyle\sum_{i=1}^{n} \dfrac{1}{1+\hat{b}t_i} - \dfrac{nt_n}{(1+\hat{b}t_n)\ln(1+\hat{b}t_n)} = 0 \end{cases} \tag{4.55}$$

These equations have to be solved numerically.

4.6. Notes and References

Software reliability is an important research area that has been studied by many researchers. Some books related to this are Musa *et al.* (1987), Xie (1991), Lyu (1996), Musa (1998) and Pham (2000). An earlier annotated biography can be found in Xie (1993). In addition, Ammar *et al.* (2000) presented a brief comparative survey of fault tolerance as it arises in hardware systems and software systems and discussed logical models as well as statistical models.

Other than Markov models discussed in this chapter, Limnios (1997) analyzed the dependability of semi-Markov systems with finite state space based on algebraic calculus within a convolution algebra. Tokuno & Yamada (2001) constructed a Markov model, which related the failure and restoration characteristics of the software system with the cumulative number of corrected faults, and also considered the imperfect debugging process together with the time-dependent property. Goseva-Popstojanova & Trivedi (2003) presented an interesting study on some architecture-based approaches in software reliability. Becker *et al.* (2000) presented a semi-Markov model for software reliability allowing for inhomogenities with respect to process time. Rajgopal & Mazumdar (2002) also presented a Markov model for the transfer of control between different software modules. Boland & Singh (2003) also investigated a birth-process approach.

For the NHPP models, Yamada & Osaki (1985) summarized some earlier software reliability growth models. Recently, many specific NHPP models have been studied. For example, Kuo *et al.* (2001) proposed a scheme for constructing software reliability growth models based on a NHPP model. Huang *et al.* (2003) further described how several existing software reliability growth models based on NHPP can be comprehensively derived by applying the concept of weighted arithmetic, weighted geometric, or weighted harmonic mean. Huang & Kuo (2003) presented some analysis that incorporates logistic testing-effort function into software reliability modeling. Zhang & Pham (2002) studied the problem of predicting operational software availability for telecommunication systems.

Shyur (2003) also presented an NHPP model that considers both imperfect debugging and the change-point. Pham (2003) recently presented studies in software reliability that includes NHPP software reliability models, NHPP models with environmental factors, and cost models. See Pham & Zhang (2003) on some further discussion on some reliability and cost models with testing coverage.

Although the Markov and NHPP models are widely used in software reliability, some other models and tools might be also useful. Miller (1986) introduced "Order Statistic" models in studying the software reliability, which can also be found in the later research of Kaufman (1996), Aki & Hirano (1996), among others. Xie *et al.* (1998) described a double exponential smoothing technique to predict software failures. Helander *et al.* (1998) presented planning models for distributing development effort among software components to facilitate cost-effective progress toward a system reliability goal. Recently, Zequeira (2000), Sahinoglu *et al.* (2001), Littlewood *et al.* (2003) and Ozekici & Soyer (2003), among others, studied some Bayesian approaches to model and estimate the reliability of software-based systems.

CHAPTER 5

MODELS FOR INTEGRATED SYSTEMS

A computing system usually integrates both software and hardware, and software cannot work without the support of hardware. Hence, computing system reliability should be studied by considering both software and hardware components. This chapter presents some models for the reliability analysis at the system level by incorporating both software and hardware failures. First, a single processor system is studied. Second, the case of modular system reliability is discussed. Following that, Markov models for clustered computing system are presented. Finally, a unified model that integrates NHPP software model into the Markov hardware model is shown.

5.1. *Single-Processor System*

The simplest case for the integrated software and hardware system is to view it as a single processor divided into two subsystems: software and hardware subsystems. Considering such system, Goel & Soenjoto (1981) presented one of the first, but general, Markov models, which will be described in this section.

5.1.1. Markov modeling

The assumptions of the model are as follows:

1) Faults in the software subsystem are independent from each other and each has a failure occurrence rate of λ.
2) Failures of hardware subsystem are also independent and have a failure occurrence rate of λ_h.
3) The time to remove a software fault, when there are i such faults in the system follows an exponential distribution with parameter μ_i.
4) The time to remove the cause of a hardware failure also follows an exponential distribution with parameter μ_h.
5) Failures and repairs of the hardware subsystem are independent of both the failures and repairs of the software subsystem.
6) At most one software fault is removed and no new software faults are introduced during the fault correction stage.
7) When the system is not operational due to the occurrence of a software failure, the fault causing the failure is corrected with probability p_s and $q_s = 1 - p_s$, is the probability of imperfect repair of software.
8) After the occurrence of a hardware failure, the hardware subsystem is recovered with probability, p_h and $q_h = 1 - p_h$ is the probability for the hardware still staying at the failed state after the repair.

Let $X(t)$ denote the state of the system at time t and '$X(t) = i$', i=0,1,...,N, implies that the system is operational while there are i remaining software faults. Here N is the initial number of software faults. Also, '$X(t) = i_s$', $i_s = 1_s, 2_s, ..., N_s$, implies that the system is down for repair of software with i remaining software faults at the time of failure. Similarly, '$X(t) = i_h$', $i_h = 1_h, 2_h, ..., N_h$, implies the system is down for repair of hardware with i remaining software faults at the time of failure. The Markov chain is shown in Fig. 5.1.

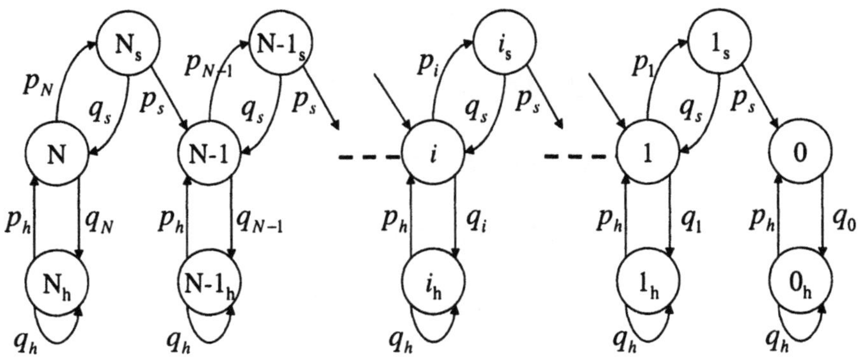

Fig. 5.1. Markov chain for the transitions between states of $X(t)$.

Suppose that the system is at state i (an operational state containing i software faults), i=1,2,...,N. The system may fail due to the software failure with probability p_i to state i_s and due to the hardware failure with the probability q_i to state i_h. At state i_s, debugging is taking place to remove the fault that causes the software failure. With probability p_s, the software fault is successfully removed and the system goes to state i-1. Otherwise with probability q_s, the fault is not removed and the software is only restarted at state i. For state i_h, maintenance personnel will try to recover the hardware failure and it has a probability p_h to return to the operational state i and probability q_h to remain at the failure state i_h. After the software is fault-free, i.e. at the state 0, the system reduces to a hardware system subject to hardware failures only.

Let $Q_{k,j}(t)$ be the one-step transition probability that, after transiting into state k, the process $X(t)$ next transits to state j in an amount of time less than or equal to t. Denoted by $F_{k,j}(t)$ the cumulative distributing function of the time from state k to state j. Then, $Q_{k,j}(t)$ is the product of $P_{k,j}$ and $F_{k,j}(t)$. The expressions for $Q_{k,j}(t)$ in the Fig. 5.1 are as follows:

$$Q_{i,i_s}(t) = p_i[1-\exp\{-(\lambda_h + i\lambda)t\}]$$

$$Q_{i,i_h}(t)=q_i[1-\exp\{-(\lambda_h+i\lambda)t\}]$$

$$Q_{i_s,i}(t)=q_s[1-\exp(-\mu_i t)]$$

$$Q_{i_s,i-1}(t)=p_s[1-\exp(-\mu_i t)] \qquad (5.1)$$

$$Q_{i_h,i}(t)=p_h[1-\exp(-\mu_h t)]$$

$$Q_{i_h,i_h}(t)=q_h[1-\exp(-\mu_h t)]$$

These basic equations describe the stochastic process as a semi-Markov process and can be used to derive some system-performance measures, see e.g. Goel & Soenjoto (1981), such as time to a specified number of software faults, system operational probabilities, system reliability and availability, and expected number of software, hardware and total failures by time t. Some of the issues are discussed in the following.

5.1.2. Time to a specified number of remaining software faults

The faults remaining in the software are sources of failures and we would like to remove them as soon as possible. However, it is not always feasible or practical to remove all of the faults during a limited time period of testing. In that case, we would like to know the distribution of time to a specified purity level, i.e., of the time to n ($0 \le n \le N$) remaining faults.

Let $T_{i,n}$ be the first passage time from state i to n, and let $G_{i,n}(t)$ be its distribution function. Consider a time interval $(r, r+dr)$. For any i, the probability of remaining in the state i, in this interval is $dQ_{i,i}(r)$, and the probability of going from the state i to i_h is $dQ_{i,i_h}(r)$.

After the process $X(t)$ reaches either state i_s or i_h, further transitions will be governed by distribution functions, $G_{i_s,n}$ and $G_{i_h,n}$, respectively. $G_{N,n}(t)$ can be obtained by taking the Laplace-Stieltjes transform of the renewal equation for $G_{i,n}(t)$, $i=n+1,\dots,N$, as shown in Goel & Soenjoto (1981):

$$G_{N,n}(t) = \phi_{n+1,N} \sum_{j=1}^{3(N-n)} \frac{\Delta_{j,N-n}}{\Theta_{3N-3n,j}} \exp(-x_j t)/x_j \tag{5.2}$$

where

$$\phi_{n+1,N} = \prod_{i=n+1}^{N} (\lambda_i p_s \mu_i)$$

$$\Delta_{j,N-n} = (-x_j + p_h \mu_h)^{N-n}$$

$$\Theta_{i,j} = \prod_{i=1}^{t} (-x_j + x_i)$$

Then, it can be shown that

$$E[T_{N,n}] = \int_0^\infty t \cdot dG_{N,n}(t) = \phi_{n+1,N} \sum_{j=1}^{3(N-n)} \frac{\Delta_{j,N-n}}{\Theta_{3N-3n,j}} (x_j)^{-2} \tag{5.3}$$

$$E[T^2{}_{N,n}] = 2\phi_{n+1,N} \sum_{j=1}^{3(N-n)} \frac{\Delta_{j,N-n}}{\Theta_{3N-3n,j}} (x_j)^{-3} \tag{5.4}$$

and

$$Var\{T_{N,n}\} = E[T^2{}_{N,n}] - E^2[T_{N,n}] \tag{5.5}$$

Example 5.1. Consider a system with $N = 10$ faults, $p_s = 0.9$ and $p_h = 0.9$. Suppose that $\lambda_i = i\lambda$, $\mu_i = i\mu$, and the parametric values are $\lambda = 0.02$, $\mu = 0.05$, $\lambda_h = 0.01$, $\mu_h = 0.025$.

Substituting the numerical values into Eq. (5.2), we obtain the distribution of $T_{N,n}$. The distribution of $T_{N,2}$ is shown in Fig. 5.2 and the trend for other distributions are similar to this. The means and standard deviations of these distributions are obtained from the above equations respectively, and some numerical results are shown in Table 5.1.

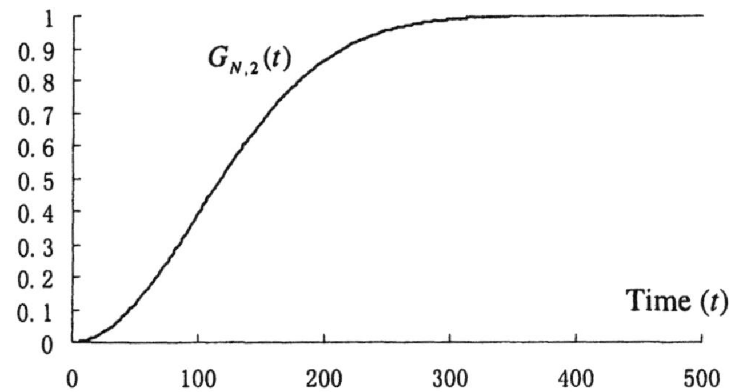

Fig. 5.2. A sample distribution of $T_{N,2}$.

Table 5.1. Mean and standard deviation of first passage-time distribution.

To State	Mean	Std. Dev.
9	10.2	17.1
8	21.6	25.0
7	34.4	31.9
6	49.1	38.4
5	66.2	45.2
4	86.7	47.7
3	112.3	61.5
2	146.4	72.9
1	197.7	90.8
0	300.1	133

■

5.1.3. System reliability and availability

The system reliability, or the probability that the system is operational at time t with a specified number of remaining software faults, can then be derived as the following. Let $P_{N,n}(t)$ be the probability that the system is operational at time t

with n remaining software faults, given that it was in operation at time $t=0$ with N software faults, i.e.,

$$P_{N,n}(t)=\Pr\{X(t)=n \mid X(0)=N\}, \quad n=0,1,\dots,N \tag{5.6}$$

We call $P_{N,n}(t)$ the (operational) state occupancy probability. By conditioning on the first up-down cycle of the process, as shown by Goel & Soenjoto (1981), the following equation for $P_{n,n}(t)$ can be obtained:

$$P_{n,n}(t)=\exp\{-(\lambda_n+\lambda_h)t\}+Q_{n,n}\otimes P_{n,n}(t) \tag{5.7}$$

In the above, $\otimes$ is the convolution operator as in Eq. (2.35). By conditioning on the first passage time, we have

$$P_{N,n}(t)=P_{n,n}(t)\otimes G_{N,n}(t) \tag{5.8}$$

where $G_{N,n}(t)$ is given by Eq. (5.2).

By taking the Laplace-Stieltjes transforms of the above equations and solving the resulting equations, we have

$$P_{N,n}(t)=G_{N,n}(t)-\phi_{n+1,N}\sum_{j=1}^{3(N-n)+3}\frac{\Delta_{j,N-n}\exp(-x_j t)}{\Theta_{3N-3n+3,j}\,x_j} \tag{5.9}$$

The system availability can then be computed as

$$A(t)=\sum_{n=0}^{N}P_{N,n}(t) \tag{5.10}$$

An example for operational probability and system availability is shown below.

Example 5.2. Continued with Example 5.1. The distributions $P_{10,n}(t)$ obtained from Eq. (5.9) and the availability function $A(t)$ obtained from Eq. (5.10) are shown in Table 5.2 for some different time points.

Table 5.2. Selected values of $P_{10,n}(t)$ and $A(t)$.

n \ t	50	100	300	500
10	.003	.001	.000	.000
9	.009	.003	.000	.000
8	.030	.006	.000	.000
7	.001	.012	.000	.000
6	.149	.024	.001	.000
5	.166	.052	.001	.000
4	.106	.106	.004	.000
3	.036	.159	.011	.001
2	.006	.139	.037	.003
1	.000	.054	.146	.033
0	.000	.005	.437	.645
A(t)	.586	.561	.637	.682

■

5.1.4. Expected number of failures by time *t*

Expected number of software failures

Let $M_s(t)$ be the expected number of software failures detected by time t. Consider a counting process $\{N_{si}(t), i \geq 0\}$, where $N_{si}(t)$ is the number of software failures detected during the time interval $(0,t]$, when the initial number of faults in the software system is i. Let $M_{si}(t) = E\{N_{si}(t) \mid X(0) = i\}$. By conditioning on the first passage time going from state N to i, we have

$$M_s(t) = M_{si}(t) \otimes G_{N,i}(t) \tag{5.11}$$

Using the Laplace-Stieltjes transforms as shown in Goel & Soenjoto (1981), we get

$$M_s(t)=\sum_{i=1}^{N}\phi_{i,N}\sum_{j=1}^{3(N-i+1)}\frac{\Delta_{j,N-i+1}(-x_j+\mu_i)}{p_s\cdot\mu_i\cdot\Theta_{3(N-i+1),j}}\cdot\frac{\exp(-x_jt)}{x_j} \tag{5.12}$$

Expected number of hardware failures

Let $M_h(t)$ be the expected number of hardware failures detected by time t. Consider a counting process $\{N_{hi}(t), i\geq 0\}$, where $N_{hi}(t)$ is the number of hardware failures detected during the time interval $(0,t]$, when the initial number of faults in the hardware subsystem is i. Let $M_{hi}(t)=E\{N_{hi}(t)\mid X(0)=i\}$. By conditioning on the first passage time from state N to i, we have

$$M_h(t)=M_{hi}(t)\otimes G_{N,i}(t) \tag{5.13}$$

Using the Laplace-Stieltjes transforms, we get

$$M_h(t)=\sum_{i=0}^{N}G_i(t) \tag{5.14}$$

The expected total number of failures denoted by $M(t)$ is the summation of software failures and hardware failures as

$$M(t)=M_s(t)+M_h(t) \tag{5.15}$$

Example 5.3. Consider the same example as in Examples 5.1 and 5.2. For this system, the expected numbers of software, hardware, and system failures are computed from the above equations. Some numerical values are given in Table 5.3.

Table 5.3 shows that the number of software failures detected increases rapidly at the beginning, leveling off at a value of about 11 at t=500. This happens because the software failure rate depends on the number of remaining faults and this number decreases with time. After t=800, there are no software faults left and the system is composed of a perfect software subsystem and a failure-prone hardware subsystem. The rate of occurrence of hardware failures, on the other hand, is unaffected by the passage of time.

Table 5.3. Expected cumulative number of failure detected.

Time	Software failures	Hardware failures	Total failures
0	0	0	0
20	2.48	0.14	2.62
40	4.19	0.26	4.45
60	5.47	0.38	5.85
80	6.47	0.49	6.96
100	7.27	0.61	7.87
200	9.59	1.17	10.77
400	10.88	2.44	13.32
600	11.08	3.80	14.88
800	11.11	5.18	16.29
1000	11.11	6.57	17.68

■

5.2. Models for Modular System

Similar to the case of modular software presented in the previous chapter, integrated software and hardware systems can also be decomposed into a finite number of modules. Markov models can also be used in analyzing such modular systems as shown below.

5.2.1. Markov modeling

Siegrist (1988) might be one of the first models using Markov processes to analyze the modular software/hardware systems. It was assumed that the control of the system is transferred among the modules according to a Markov process. Each module has an associated reliability which gives the probability that the module will operate correctly when called and will transfer control successfully when finished. The system will eventually either fail or complete its task successfully so that to enter a terminal state.

The modules (or states) of the system is denoted by i (i=1,2,...,n). The ideal (failure free) system is described by a Markov chain with state space $\{1,2,\ldots,n\}$ and transition matrix **P**. That is, P_{ij} is the conditional probability that the next state will be j given that the current state is i. The reliability of state i, denoted by R_i, is the probability that state i will function correctly when called and will transfer control successfully when finished. The imperfect system is modeled by adding an absorbing state F (failure state) and the transition matrix is modified accordingly.

Specifically, the imperfect system is described by a Markov chain with state space $\{1,2,\ldots,n,F\}$ and transition matrix $\hat{P}$ given by

$$\hat{P}_{ij} = R_i P_{ij}, \quad \text{for } i, j=1,\ldots,n$$

$$\hat{P}_{iF} = 1 - R_i, \quad \text{for } i=1,\ldots,n \tag{5.16}$$

$$\hat{P}_{FF} = 1$$

Usually $R_i < 1$ for each i and hence each of the states $1,2,\ldots,n$ eventually leads to the absorbing state F. Note that the dynamics of the imperfect system are completely described by the state reliability function $\boldsymbol{R}$ and the transition matrix $\boldsymbol{P}$ since this description is equivalent to specifying the transition matrix $\hat{P}$ of the imperfect system.

5.2.2. Expected number of transitions until failure

Based on the above Markov model, Siegrist (1988) presented the expected number of transitions until failure as the measure of system reliability. Let M_i denote the expected number of transitions until failure for the imperfect system, starting in state *i*. If the transitions of the system correspond to inputs received at regular time intervals, then M_i is proportional to the expected time until failure, starting in state *i*. Two methods of computing the function *M* will be given.

Let **Q** denote the restriction of the transition matrix $\hat{P}$ of the imperfect system to the (transient) states $1,2,\ldots,n$. Note that $Q_{ij} = R_i P_{ij}$. Then

$$\sum_{k=0}^{\infty} \mathbf{Q}^k = (\mathbf{I}-\mathbf{Q})^{-1} \tag{5.17}$$

It follows that

$$M_i = (\mathbf{I}-\mathbf{Q})_{ij}^{-1} \tag{5.18}$$

Let *i* and *j* be any of the states $1,2,\ldots,n$. We have that

$$M_i = A_{ij} + B_{ij} M_j \tag{5.19}$$

where A_{ij} is the expected number of transitions until the imperfect system either fails or reaches state *j*, starting in state *i*; and B_{ij} is the probability that the imperfect system eventually reaches state *j*, starting in state *i*. If *i*=*j*, "reaches" should be interpreted as "returns to" in which case, we obtain from the above equation

$$M_j = \frac{A_{jj}}{1-B_{ij}} \tag{5.20}$$

Then, the desired result is

$$M_i = A_{ij} + \frac{A_{jj} B_{ij}}{1-B_{ij}} \tag{5.21}$$

From the Markov property, the matrices **A** and **B** are related to the basic data **R** and **P** according to the following systems of equations:

$$A_{ij} = 1 + \sum_{k \neq j} R_i P_{ik} A_{kj} \tag{5.22}$$

and

$$B_{ij} = R_i P_{ij} + \sum_{k \neq j} R_i P_{ik} B_{kj} \tag{5.23}$$

Moreover $B_{ij}(1-B_{ij})$ is the same as $(\mathbf{I}-\mathbf{Q})_{ij}^{-1}$, namely, the expected number of visits to j for the imperfect system starting in state i.

With the Markov property, the measure of expected number of transitions until failure is derived. As a result, this model is more appropriate for systems which run for fixed periods of time or which run continuously (until failure). Two examples of branching and sequential structures are illustrated here.

Example 5.4. (A Branching System) A general branching system has the transition graph depicted in Fig. 5.3. State 1 acts as a central control which may pass the control to any of the states 2,...,n or back to itself. Each of the branch states 2,...,n can pass control back to itself or back to the center state 1.

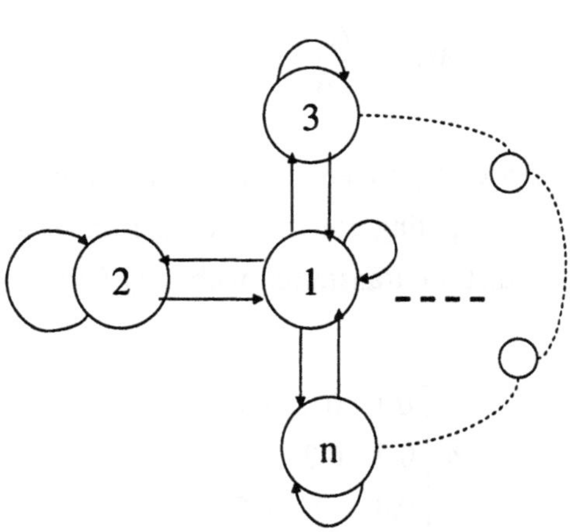

Fig. 5.3. The transition graph for a general branching system.

Given the transition matrix **P** and the state reliability function **R** and let state 1 be the initial state, M_1, the expected number of transitions until failure starting in state 1, can be computed. Note first that the imperfect system, starting in state 1 will make at least one transition before failure or return to state 1 occurs. Furthermore, if the stem moves to state *j* on the first transition, then on average, the system will make $1/(1-R_jP_{jj})$ transitions until failure or return to state 1 occurs. It follows that

$$A_{11} = 1 + \sum_{j=2}^{n} \frac{R_1 P_{1j}}{1 - R_j P_{jj}} \tag{5.24}$$

On the other hand, the probability that the imperfect system, starting from state 1, will eventually return to state 1 is

$$B_{11} = R_1 P_{11} + \sum_{j=2}^{n} \frac{R_1 P_{1j} R_j P_{j1}}{1 - R_j P_{jj}} \tag{5.25}$$

Therefore, from Eq. (5.20)

$$M_1 = \frac{A_{11}}{1 - B_{11}} \tag{5.26}$$

If n=3 modules including a CPU, a memory and a computing software, CPU is the central state that any computing control starting from it and the other two modules are branch states that are transferred with only CPU and itself. Given a transition matrix

$$\mathbf{P} = \begin{bmatrix} 0.1 & 0.3 & 0.6 \\ 0.7 & 0.3 & 0 \\ 0.8 & 0 & 0.2 \end{bmatrix}$$

and reliability $R_1 = 0.95$, $R_2 = 0.9$, $R_3 = 0.85$, and by substituting the numerical values into the above equations, we get

$$A_{11} = 2.077 \quad \text{and} \quad B_{11} = 0.8079$$

The expected number of transition till failure is

$$M_1 = 10.815$$

Given the expected time of each transition is 26 seconds, the MTTF is

$$MTTF = 10.815 \times 26 = 281.2 \quad \text{(seconds)}$$

■

Example 5.5. (A Sequential System) The transition graph of a sequential system is given in Fig. 5.4. Note that control tends to pass sequentially from state 1 or state 2,..., to state n except that in each state, control can return to that state or to state 1 which is the initial state.

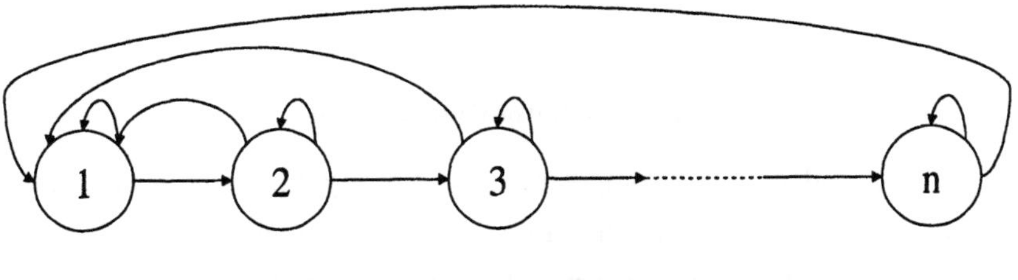

Fig. 5.4. The transition graph of a sequential system.

Suppose that the transition matrix **P** and the state reliability function **R** are known. First note that when the system is in state i, the expected number of transitions until the process leaves state i is $1/(1 - R_i P_{ii})$. It follows that

$$A_{11} = 1 + \frac{R_1 P_{12}}{1 - R_2 P_{22}} + R_1 P_{12} \sum_{i=3}^{n} \frac{\prod_{k=2}^{i-1} \frac{R_k P_{k,k+1}}{1 - R_k P_{kk}}}{1 - R_i P_{ii}} \tag{5.27}$$

By a similar argument, the probability of eventual return to state 1, starting in state 1 for the imperfect system is

$$B_{11} = R_1P_{11} + \frac{R_1P_{12}R_2P_{21}}{1-R_2P_{22}} + R_1P_{12}\sum_{i=3}^{n}\left\{\left[\prod_{k=2}^{i-1}\frac{R_kP_{k,k+1}}{1-R_kP_{kk}}\right]\cdot\frac{R_iP_{i1}}{1-R_iP_{ii}}\right\} \tag{5.28}$$

■

5.3. Models for Clustered System

Clustered computing systems use commercially available computers networked in a loosely-coupled fashion. It can provide high levels of reliability if appropriate levels of fault detection and recovery software are implemented in the middleware (an application layer). The application, therefore, can be made as reliable as the user requires and it is constrained only by the upper bounds on reliability imposed by the architecture, performance and cost considerations.

5.3.1. Introduction to clustered computing systems

A cluster is a collection of computers in which any member of the cluster is capable of supporting the processing functions of any other member. A clustered computing system has a redundant $n+k$ configuration, where n processing nodes are actively processing the application and k processing nodes are in a standby state, serving as spares. In the event of a failure of an active node, the application that was running on the failed node is moved to one of the standby nodes.

The simplest cluster system is one *active* and one *standby*, in which one node is actively processing the application and the other node is in a standby state. Other common cluster systems include *simplex* (one active node, no spare), n+1 (n active nodes, 1 spare), and n+0 (all n active nodes). In a system with n active nodes, the applications from the failed node are redistributed among the other active nodes using a pre-specified algorithm.

Consider a general clustered computing system with n active processors and k spares, see e.g., Mendiratta (1998). In this system, there is a Power Dog (PD) attached to each processor that can *power cycle* or *power down* the processor, and a Watch Dog (WD) with connections to each processor that monitors performance from each processor and initiate *failover* if it detects a processor failure. Then, the failover information is transferred to a switching system (SS) that can turn on the Power Dog of the standby processors to replace the failed ones.

The block diagram for this clustered system architecture is shown in Fig. 5.5 and represents the system to be modeled.

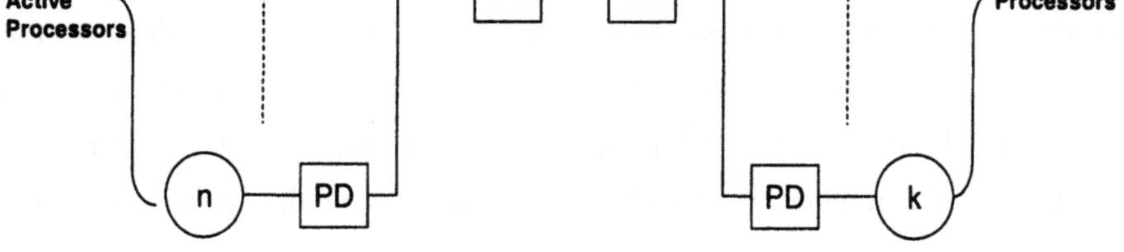

Fig. 5.5. A general architecture of $n+k$ clustered computing systems.

5.3.2. Markov modeling

For each processor, there are two types of failures: software and hardware failures. Suppose the failure rate for software is λ_s and for hardware λ_h. Those failed processors may or may not be repaired, which will be discussed in the following, respectively.

Model for non-repairable system

Non-repairable system means that the processors are not repaired if they are failed. Thus, for the $n+k$ clustered system without repair, the Markov model can be depicted by the CTMC in Fig. 5.6.

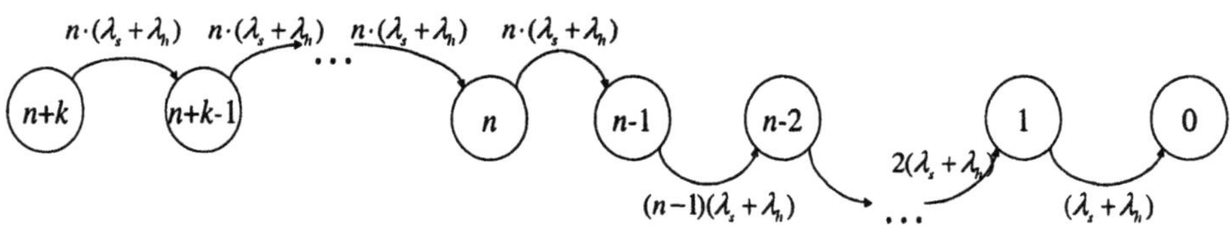

Fig. 5.6. CTMC of $n+k$ clustered system without repair.

The state i in Fig. 5.6 represents the number of good processors (both active and standby). If $i \geq n$, the cluster system must keep n processors active, so the failure rate should be $n(\lambda_s+\lambda_h)$. If $0<i<n$, it means that no spares are available and the number of active processors is i. Hence, the failure occurrence rate is $i(\lambda_s+\lambda_h)$.

Denote by $P_i(t)$ (i=0,1,2,..., $n+k$) the probability for the system to stay at state i at time instant t. The Chapman-Kolmogorov equation can be written as

$$P_{n+k}'(t) = -n(\lambda_s+\lambda_h)P_{n+k}(t)$$

$$P_i'(t) = n(\lambda_s+\lambda_h)P_{i+1}(t) - n(\lambda_s+\lambda_h)P_i(t), \quad i = n, n+1, ..., n+k-1$$

$$P_i'(t) = (i+1)(\lambda_s+\lambda_h)P_{i+1}(t) - i(\lambda_s+\lambda_h)P_i(t), \quad i=1,2,3,...,n-1$$

$$P_0'(t) = (\lambda_s+\lambda_h)P_1(t) \tag{5.29}$$

We assume that the process begins from the state *n+k* that all the processors are good initially. Hence, the initial conditions are

$$P_{n+k}(0)=1,\text{ and } P_{n+k-1}(0)=P_{n+k-2}(0)=\ldots=P_0(0)=0 \tag{5.30}$$

With a numerical program, one can obtain the solution of the above differential equations with initial conditions even for large value of *n+k*.

The probability of the system failure state $P_0(t)$ determines the unreliability function. Therefore, the reliability function defined as the probability that at least one processor works well is

$$R(t)=1-P_0(t) \tag{5.31}$$

Moreover, we can use Laplace-Stieltjes transform to approximate the reliability function. For example, the state probability for the failed state after the transformation is

$$P_0(s)=\frac{\lambda_1\lambda_2\ldots\lambda_{n+k}}{(s+\lambda_1)(s+\lambda_2)\ldots(s+\lambda_{n+k})s} \tag{5.32}$$

where

$$\lambda_i=\begin{cases} n(\lambda_s+\lambda_h) & \text{if } i\ge n \\ i(\lambda_s+\lambda_h) & \text{if } i<n \end{cases}$$

Expanding the denominator, substituting the expression in the equation for $P_0(s)$, we obtain:

$$P_0(s)=\frac{1}{s^{n+k+1}}\prod_{i=1}^{n+k}\lambda_i-\frac{1}{s^{n+k}}\prod_{i=1}^{n+k}\lambda_i\sum_{i=1}^{n+k}\lambda_i+\ldots \tag{5.33}$$

The above equation can be easily inverted using inverse Laplace-Stieltjes transforms

$$P_0(t)=\frac{t^{n+k}}{(n+k)!}\prod_{i=1}^{n+k}\lambda_i-\frac{t^{n+k-1}}{(n+k-1)!}\prod_{i=1}^{n+k}\lambda_i\sum_{i=1}^{n+k}\lambda_i+\ldots \tag{5.34}$$

Generally, we use only the first term for the approximation and we have

$$R(t) \approx 1 - \frac{t^{n+k}}{(n+k)!}\prod_{i=1}^{n+k}\lambda_i \tag{5.35}$$

Since we have an alternating power series, the next term will provide a bound on the absolute error in using this approximation:

$$|Error| < \frac{t^{n+k-1}}{(n+k-1)!}\prod_{i=1}^{n+k}\lambda_i \sum_{i=1}^{n+k}\lambda_i$$

Model for repairable system

If a system is repairable, the failed processor can be recovered with a repair rate μ_i from state *i*-1 back to state *i*. The Markov model is built as the CTMC of Fig. 5.7.

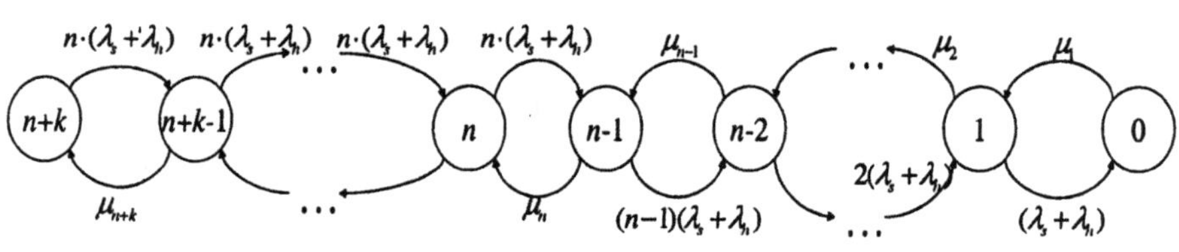

Fig. 5.7. CTMC for repairable clustered systems.

As before, the Chapman-Kolmogorov equation can be written as

$$P_{n+k}'(t) = \mu_{n+k}P_{n+k-1}(t) - n(\lambda_s + \lambda_h)P_{n+k}(t)$$

$$P_i'(t) = n(\lambda_s + \lambda_h)P_{i+1}(t) + \mu_i P_{i-1}(t) - [n(\lambda_s + \lambda_h) + \mu_{i+1}]P_i(t)\text{, } i = n,\ldots,n+k-1$$

$$P_i'(t) = (i+1)(\lambda_s + \lambda_h)P_{i+1}(t) + \mu_i P_{i-1}(t) - [i(\lambda_s + \lambda_h) + \mu_{i+1}]P_i(t),\ i=1,\ldots,n\text{-}1$$

$$P_0'(t) = (\lambda_s + \lambda_h)P_1(t) - \mu_1 P_0(t) \tag{5.36}$$

with the initial conditions (5.30).

Again, these equations can be solved numerically using certain computer programs.

Model with different repair rates of software and hardware

In the above model for repairable clusters, μ_i is the expected system repair rate no matter whether the failed processors are caused by software failures or hardware failures. Actually, the rate for repairing software failure should be different from that for repairing hardware failure (Lai *et al.*, 2002). A model for this different repair rates is discussed here.

Let μ_s be the rate to repair one failed processor caused by software failure and μ_h by hardware failure. Then part of the CTMC can be depicted as shown in Fig. 5.8.

In Fig. 5.8, the transition rates are given by

$$\lambda_s(i,j) = \begin{cases} n\lambda_s & \text{if } i+j \le k \\ (n-i-j)\lambda_s & \text{if } i+j > k \end{cases}$$

and

$$\lambda_h(i,j) = \begin{cases} n\lambda_h & \text{if } i+j \le k \\ (n-i-j)\lambda_h & \text{if } i+j > k \end{cases}$$

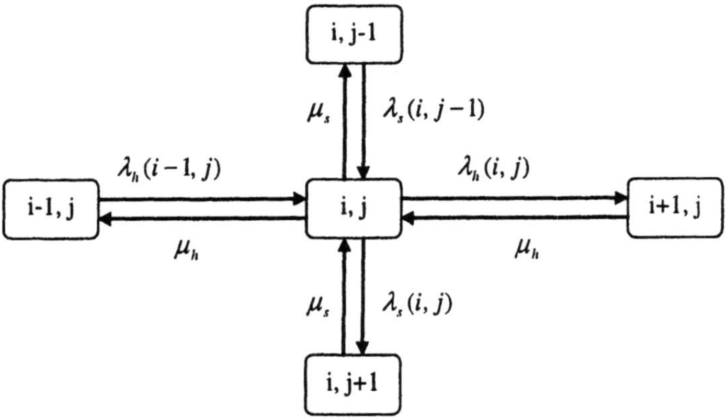

State(i,j): i hardware down, j software down (on different processors)

Fig. 5.8. CTMC for repairable cluster with different software/hardware repair rate.

The corresponding Chapman-Kolmogorov differential equation for the probability that the system is in the state (*i*, *j*) at time *t* is, for $i, j \neq 0, n+k;\ i+j \leq n+k-1$,

$$
\begin{aligned}
P_{i,j}'(t) &= \mu_h P_{i+1,j}(t) + \lambda_h(i-1, j)P_{i-1,j}(t) + \lambda_s(i, j-1)P_{i,j-1}(t) \\
&+ \mu_s P_{i,j+1}(t) - [\mu_s + \lambda_h(i, j) + \lambda_s(i, j) + \mu_h]P_{i,j}(t)
\end{aligned}
\tag{5.37}
$$

The initial conditions are

$$P_{0,0}(0) = 1 \text{ and } P_{i,j}(0) = 0, \textit{ for } i, j \neq 0 \tag{5.38}$$

The boundary conditions are:

$$P_{0,0}'(t) = \mu_h P_{1,0}(t) + \mu_s P_{0,1}(t) - n(\lambda_s + \lambda_h)P_{0,0}(t)$$

$$P'_{0,j}(t) = \mu_h P_{1,j}(t) + \lambda_s(0, j-1)P_{0,j-1}(t) + \mu_s P_{0,j+1}(t)$$
$$-[\mu_h + \mu_s + \lambda_s(0, j) + \lambda_h(0, j)]P_{0,j}(t) \qquad \text{for } j = 1,2,\ldots,n+k-1$$

$$P'_{i,0}(t) = \mu_h P_{i+1,0}(t) + \lambda_h(i-1,0)P_{i-1,0}(t) + \mu_s P_{i,1}(t)$$
$$-[\mu_h + \mu_s + \lambda_s(i,0) + \lambda_h(i,0)]P_{i,0}(t) \qquad \text{for } i = 1,2,\ldots,n+k-1$$

$$P'_{i,j}(t) = \lambda_h(i-1, j)P_{i-1,j}(t) + \lambda_s(i, j-1)P_{i,j-1}(t)$$
$$-(\mu_s + \mu_h)P_{i,j}(t) \qquad \text{for } i + j = n + k; 0 < i, j < n + k$$

$$P'_{n+k,0}(t) = \lambda_h P_{n+k-1,0}(t) - \mu_h P_{n+k,0}(t) \tag{5.39}$$

and

$$P'_{0,n+k}(t) = \lambda_s(t)P_{0,n+k-1}(t) - \mu_s P_{0,n+k}(t)$$

The above equations need to be solved numerically with a computer program. After that, the system availability for the $n+k$ clustered system can be calculated by

$$A(t) = \sum_{i+j<n+k} P_{i,j}(t) \tag{5.40}$$

Example 5.6. Consider a clustered system containing 2 active processors. Suppose that the failure rate of software is $\lambda_s = 0.003$ and that of hardware is $\lambda_h = 0.0012$ in a processor. We discuss the following three different conditions in the following.

1) The case without repair

In this case, the Markov model for this 2+0 cluster is constructed in Fig. 5.9.

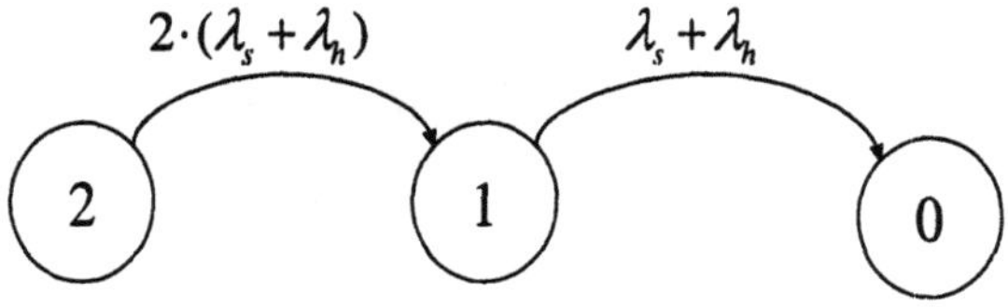

Fig. 5.9. CTMC without considering repair.

Solving the Chapman-Kolmogorov differential equations in (5.28) with initial condition (5.29), we get

$$P_0(t) = [1 - \exp\{-(\lambda_s + \lambda_h)t\}]^2$$

and the reliability function

$$R(t) = 1 - P_0(t) = 1 - [1 - \exp(-0.0042t)]^2$$

2) The case of the identical repair rate

With the identical repair rate μ, the Markov model for this 2+0 cluster is constructed in Fig. 5.10.

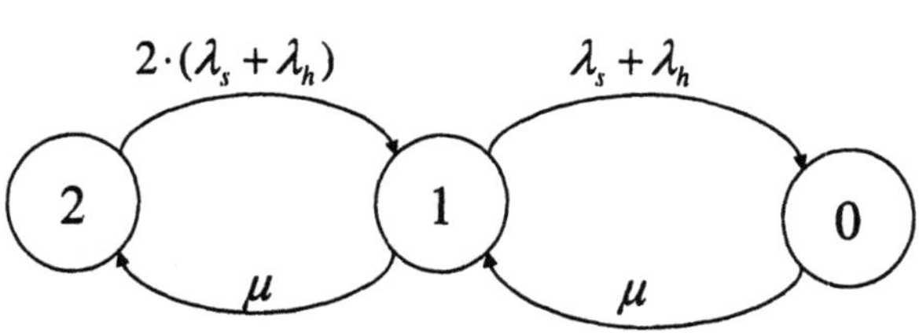

Fig. 5.10. CTMC for Example 5.6 considering identical repair rate.

By constructing the Chapman-Kolmogorov differential equations, we get

$$P_2'(t) = \mu P_1(t) - 2(\lambda_s + \lambda_h)P_2(t)$$

$$P_1'(t) = 2(\lambda_s + \lambda_h)P_2(t) + \mu P_0(t) - (\lambda_s + \lambda_h + \mu)P_1(t)$$

$$P_0'(t) = (\lambda_s + \lambda_h)P_1(t) - \mu P_0(t)$$

with the initial condition $P_2(0) = 1, P_1(0) = P_0(0) = 0$. If the repair rate $\mu = 0.01$ for both software failure and hardware failure, we can obtain the availability function $A(t) = 1 - P_0(t)$ numerically, as shown in Fig. 5.11.

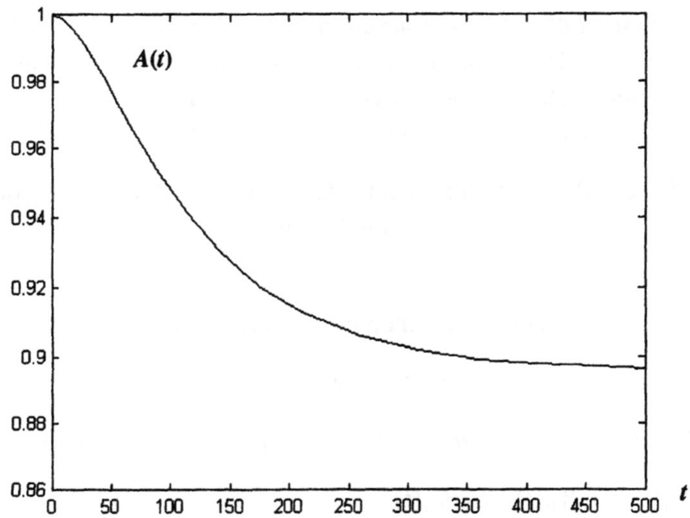

Fig. 5.11. Availability function considering identical software/hardware repair rate.

3) General case of different repair rates

Considering different repair rates for software and hardware, the Markov model for this 2+0 cluster is constructed in Fig. 5.12.

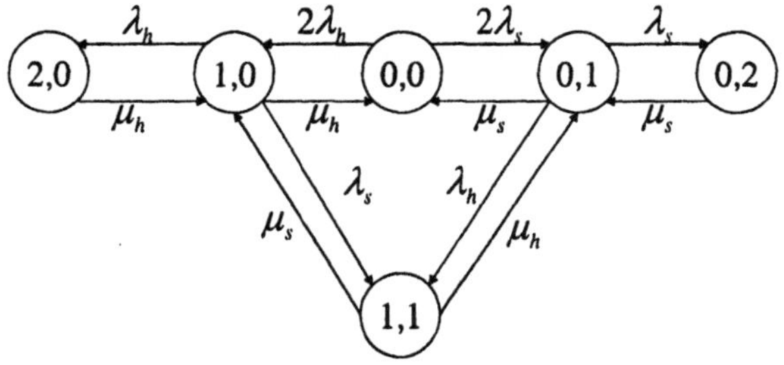

State (0,0): initial state, all components working
State (1,0): 1 hardware down, 1 host working
State (2,0): 2 hardware down, system down
State (1,1): 1 hardware down, 1 software down, system down
State (0,1): 1 software down, 1 host working
State (0,2): 2 software down, system down

Fig. 5.12. CTMC for Example 5.6 with different software and hardware repair rate.

The corresponding Chapman-Kolmogorov differential equations are

$$P_{0,0}'(t) = \mu_h P_{1,0}(t) + \mu_s P_{0,1}(t) - (2\lambda_h + 2\lambda_s) P_{0,0}(t)$$
$$P_{1,0}'(t) = 2\mu_h P_{0,0}(t) + \mu_h P_{2,0}(t) + \mu_s P_{1,1}(t) - (\mu_h + \lambda_h + \lambda_s) P_{1,0}(t)$$
$$P_{2,0}'(t) = \lambda_h P_{1,0}(t) - \mu_h P_{2,0}(t) \tag{5.41}$$
$$P_{1,1}'(t) = \lambda_s P_{1,0}(t) + \mu_h P_{0,1}(t) - (\mu_h + \mu_s) P_{1,1}(t)$$
$$P_{0,1}'(t) = 2\lambda_h P_{0,0}(t) + \mu_h P_{1,1}(t) + \mu_s P_{0,2}(t) - (\mu_s + \lambda_h + \lambda_s) P_{0,1}(t)$$
$$P_{0,2}'(t) = \lambda_s P_{0,1}(t) - \mu_s P_{0,2}(t)$$

with initial condition $P_{0,0}(0) = 1$ and other probabilities 0. If the repair rates are $\mu_s = 0.008$ and $\mu_h = 0.012$, the availability function $A(t) = 1 - P_0(t)$ can be computed numerically.

■

5.4. A Unified NHPP Markov Model

In order to incorporate the NHPP software reliability model into the Markov hardware reliability model, Welke *et al.* (1995) developed a unified NHPP Markov model. The unified model is accomplished by determining a transition probability for a software failure and then incorporating the software failure transitions into the hardware reliability model. Based on this unified model, the differential equations can be easily established and solved despite the time-varying software failure rates.

The basic assumptions of this unified model are listed below:

1) Software failures are described by a general NHPP model, with the probability function

$$P(n,t)=\Pr\{N(t)=n\}=\frac{[m(t)]^n}{n!}\exp\{-m(t)\},\ n=0,1,2,\ldots \tag{5.42}$$

where $m(t)$ is the mean value function and n is the number of failures occurring up to time t.

2) The times between hardware failures are exponentially distributed random variable.

5.4.1. Software failure transition probability

The mathematical justification for implementing the NHPP model as a Markov process is based on the concept of r:th order inter-arrival times (Drake, 1967). Denoted by l_r the random variable of the r:th order inter-arrival times and let $f_{l_r}(l)$ be the probability density function of l_r we then have

$$\Pr\{l<l_r<l+\Delta l\}\approx\int_l^{l+\Delta l} f_{l_r}(\xi)d\xi=f_{l_r}(l)\Delta l \tag{5.43}$$

This equation provides a discrete-time relationship that can be incorporated into the discrete-time Markov model. The time between failures in a Poisson model has an exponential distribution, so the same derivation used in hardware modeling can be used here to show:

$$\Pr\{\text{failure occurs in } (t, t+\Delta t]\} \approx \lambda \Delta t \tag{5.44}$$

Given this approximation, the discrete-time relationship can be written in a slightly different form as:

$$f_{l_r}(l)\Delta t = P(r-1, l)\lambda \Delta l \tag{5.45}$$

where $P(r-1, l)$ is the probability that there are exactly $r-1$ failures in an interval of duration l.

Note that the failure intensity of NHPP is time-varying (Welke *et al.*, 1995) and Eq. (5.45) becomes:

$$f_{l_r}(l)\Delta t = P(r-1, l)\lambda(t)\Delta l \tag{5.46}$$

Substituting Eq. (5.42) into the above equation, we have

$$f_{l_r}(l)\Delta t = \frac{[m(l)]^{r-1}}{(r-1)!}\exp\{-m(l)\}\lambda(t)\Delta l \tag{5.47}$$

5.4.2. Markov modeling

We now use the above equation to describe software state transitions in a Markov model. The transition we evaluate is the probability that the software remains in the same (operational) state, given it started in the state. Since Eq. (5.43) gives the probability that failure r occurs in $[l, l+\Delta l]$, the probability that any software failure occurs in this interval is simply the sum of Eq. (5.43) over all possible values of r. Assume that the maximum value of r is large enough to approximate this sum as

$$\sum_{r=1}^{\infty} f_{l_r}(l)\Delta l \tag{5.48}$$

Therefore, the probability that no failures occur in $[l, l+\Delta l]$ is

$$f_{l_0}(l)\Delta l = 1 - \sum_{r=1}^{\infty} f_{l_r}(l)\Delta l \tag{5.49}$$

By substituting Eq. (5.47) into the above equation and performing some algebraic manipulation, we have

$$f_{l_0}(l)\Delta l = 1 - \lambda(t)\Delta l \sum_{r=1}^{\infty} \left\{ \frac{[m(l)]^{r-1}}{(r-1)!} \exp\{-m(l)\} \right\}$$

$$= 1 - \lambda(t)\Delta t \tag{5.50}$$

since $\Delta l = \Delta t$. The above equation means the probability that no software failure occurs during a short time Δt, so the transition probability from the operational state to the software failure state during the short enough Δt can be expressed as

$$\Pr\{\text{software failure occurs in } (t, t+\Delta t]\} = \lambda(t)\Delta t \tag{5.51}$$

Hence, with the above transition probability, NHPP model can be integrated into the Markov model. For details, see Welke *et al.* (1995). Based on the above equations, the differential equations can be obtained and solved as usual. An example for it is illustrated below.

Example 5.7. Suppose that a processing element contains both software and hardware parts. The software failures follow a classical NHPP model, the GO-model (Goel & Okumoto, 1979) with failure intensity function

$$\lambda_s(t) = ab\exp(-bt)$$

and of the hardware follow the exponential distribution with parameter λ_h. Suppose $a = 0.001$, $b = 10$, $\lambda_h = 0.0005$ and the failed system will be repaired with repair rate $\mu_s = 0.01$ for software and $\mu_h = 0.02$ for hardware.

The state probabilities satisfy the following differential equations

$$P_0'(t) = P_2(t)\mu_h + P_1(t)\mu_s - P_0(t)[\lambda_s(t) + \lambda_h]$$

$$P_1'(t) = P_0(t)\lambda_s(t) - P_1(t)\mu_s$$

$$P_2'(t) = P_0(t)\lambda_h - P_2(t)\mu_h$$

with the initial condition $P_0(0) = 1, P_1(0) = P_2(0) = 0$, we can get the availability function $A(t) = P_0(t)$ numerically, as shown by Fig. 5.13.

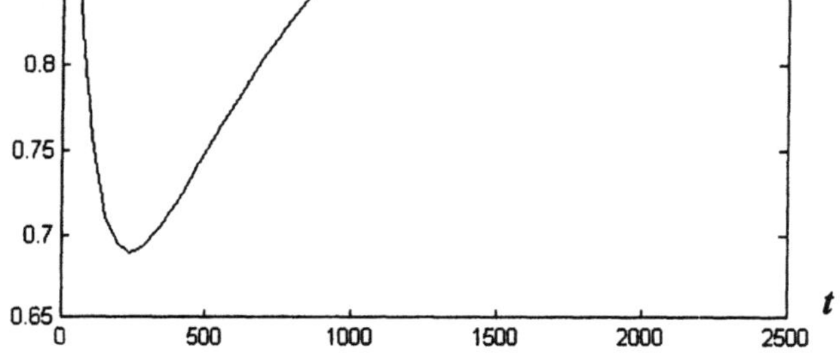

Fig. 5.13. Availability function for Example 5.7.

■

Note that although this example implemented the GO model for software failures, other NHPP software models, see e.g. Xie (1991), can also be integrated into the unified model according to their specific conditions.

5.5. Notes and References

Pukite & Pukite (1998) summarized some simple models for the reliability analysis of the hardware and software system. Another useful reference is Kapur *et al.* (1998).

Similar to the single-processor model presented in this chapter, Hecht & Hecht (1986) also studied the reliability in the system context considering both software and hardware. Fryer (1985) implemented the fault tree analysis in analyzing the reliability of combined software/hardware systems, which determines how component failures can contribute to system failure. Sumita & Masuda (1986) developed a combined hardware/software reliability model where both lifetimes and repair times of software and hardware subsystems are considered together. Kim & Welch (1989) examined the concept of distributed execution of recovery blocks as an approach for uniform treatment of hardware and software faults. Keene & Lane (1992) reviewed the similarities and differences between hardware, software and system reliability. Kanoun & Ortalo-Borrel (2000) explicitly modelled the case of hardware and software component-interactions.

For the clustered systems, Laprie & Kanoun (1992) presented Markov models for analyzing the system availability. Later, Dugan & Lyu (1995) discussed the modeling and analysis of three major architectures of the clustered system containing multiple versions of software/hardware, and they combined fault tree analysis techniques and Markov modeling techniques to incorporate transient and permanent hardware faults as well as unrelated and related software faults.

Recently, Pasquini *et al.* (2001) considered the reliability for systems based on software and human resources. Choi & Seong (2001) studied a system considering software masking effects on hardware faults. Zhang & Horigome (2001) discussed the availability and reliability on the system level considering the time-varying failures that are dependent among the software/hardware components. Lai *et al.* (2002) studied the reliability of the distributed software/hardware systems, where Markov models were implemented by assuming that the software failure rate is decreasing while the hardware has a constant failure rate. Dai *et al.* (2003a) further studied the reliability and availability of distributed services which combined both software program failures and hardware network failures altogether.

CHAPTER

6

AVAILABILITY AND RELIABILITY OF DISTRIBUTED COMPUTING SYSTEMS

Distributed computing system is a type of widely-used computing system. The performance of a distributed computing system is determined not only by software or hardware reliability but also by the reliability of networks for communication. This chapter presents some results on the availability and reliability of distributed computing systems by considering the failures of software programs, hardware processors and network communication. Graph theory and Markov models are mainly used.

The chapter is divided into four parts. First, general distributed computing system and some specific commonly used systems are introduced. Second, the distributed program/system reliability is analyzed and some analytical tools of evaluating them are demonstrated. The homogeneous distributed software/hardware system is then studied. The system availability is analyzed by Markov models and the imperfect debugging process is further introduced. Finally, the Centralized Heterogeneous Distributed System is studied and approaches to its service reliability are shown.

6.1. Introduction to Distributed Computing

The distributed computing system is designed to complete certain computing tasks given a networked environment, e.g. Casavant & Singhal (1994) and Loy *et al.* (2001). Such systems have gained in popularity due to the low-cost processors in the recent years. A common distributed system is made up of several hosts connected by a network where computing functions are shared among the hosts, as depicted by Fig. 6.1.

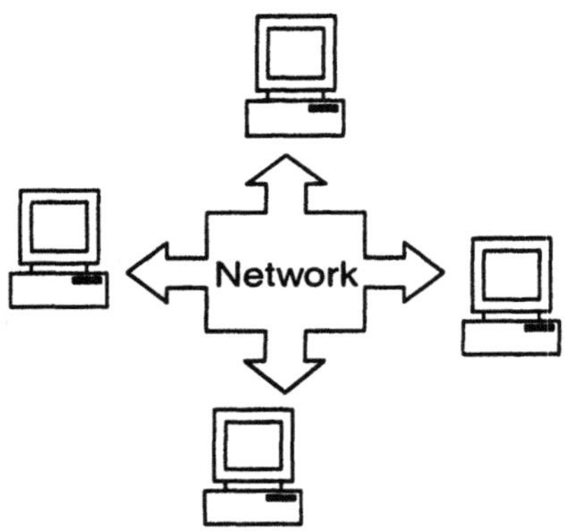

Fig. 6.1. General distributed computing system.

A typical application in distributed systems is distributed software of which identical copies run on all the distributed hosts. A homogeneously distributed system is a system for which all of the distributed hosts are of the same type, such as workstations from the same vendor. Applications of identical copies of distributed software to homogeneously distributed systems are called *homogeneously distributed software/hardware systems* (Lai *et al.*, 2002).

For example, a search engine system provides the service for searching related information. To receive and serve millions of searching requests everyday, the search engine system should contain many servers of the same type running the identical software in exploring the database. Such system is a type of the homogeneous distributed software/hardware system. Examples of applications of this kind of systems can also be found in communication protocols, telephone switching systems, web services, and distributed database management systems, etc.

Besides the homogeneous distributed systems, most of the other distributed systems can be attributed to *centralized heterogeneous distributed system* (Dai *et al.* 2003a). This kind of system consists of many heterogeneous subsystems managed by a control center.

For example, in modern warfare, each soldier can be considered as an element in a military system and furnished with different electrical equipments for diverse purposes. The information collected from each soldier is sent back to a control center through wireless communication channels. Then, the control center can analyze all the information and send out commands to respective soldiers. The functions of different groups of soldiers are diversified in a war (such as attacking, defending, supplying, saving etc.) so their electrical equipments should also be heterogeneous. Thus, it is a typical Centralized Heterogeneous Distributed System, as depicted by Fig. 6.2.

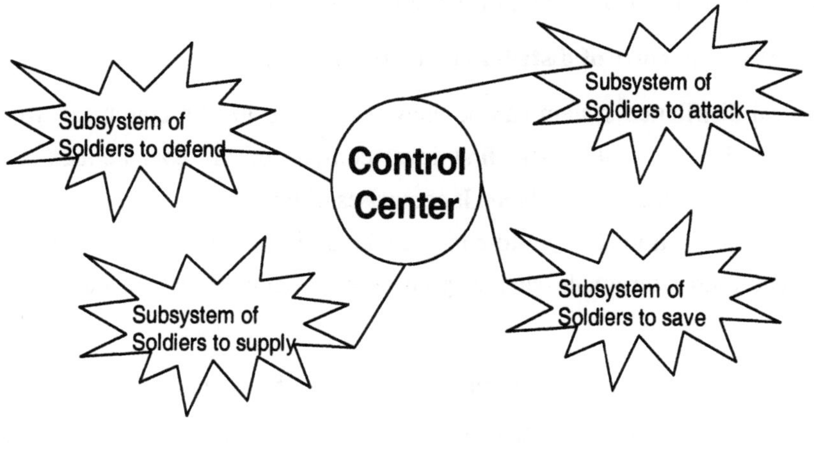

Fig. 6.2. A military system.

The reliability of distributed system is a key point of the QoS (Quality of Service). However, reliability analysis of such systems is complicated due to its various topologies, the integrated software and hardware or highly heterogeneous subsystems. This chapter studies these issues and presents models and analytical tools that can be easily implemented to estimate the reliability and availability of those distributed systems.

6.2. Distributed Program and System Reliability

A general distributed computing system consists of processing elements (nodes), communication channels (links), memory units, data files, and programs. These resources are interconnected via a network that indicates how information flows among them. Programs residing on some nodes can use/load data files from other nodes. Hence, the program/system reliability in the general networked environment is worth studying in order to comprehensively qualify the distributed system.

6.2.1. Architecture and reliability model of distributed systems

General architecture of distributed computing systems

A typical distributed system can be viewed as a two level hierarchical structure (Pierre & Hoang, 1990). The first level consists of the *communication sub-network*, also called the *backbone*. It comprises of linked switching nodes and has as its main function the end-to-end transfer of information. The second level consists of nodes/terminals, such as processors, programs, files, resources and so on.

In general, n-processor distributed systems can be depicted as Fig. 6.3. Each node can execute a set of programs PN_i and share a set of data files FN_i

(i=1,2,...,n). Programs residing on some nodes can be run using data files at other nodes.

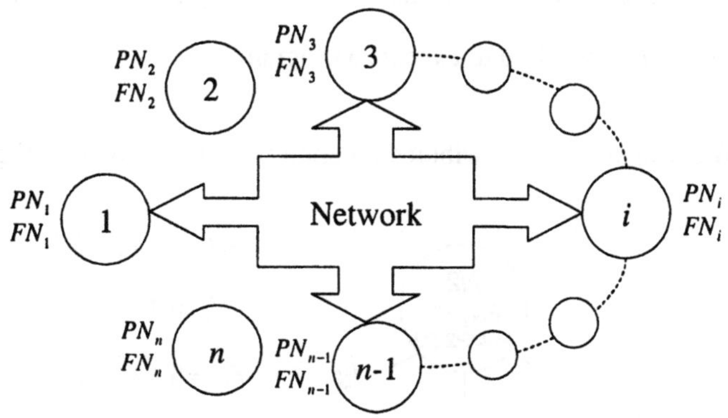

Fig. 6.3. *n* processors of a distributed computing system.

Reliability model

Based on the above model for the general distributed computing systems, the definition of the distributed program reliability is given below:

Definition 6.1. *Distributed program reliability* in a distributed computing system is the probability of successful execution of a program running on multiple processing elements and needs to retrieve data files from other processing elements.

From the definition, the distributed program reliability varies according to

1) the network topology of the distributed computing system

2) the reliability of the communication links
3) the reliability of the processing nodes
4) the data files and programs distribution among processing elements
5) the data files required to execute a program.

Example 6.1. Consider the distributed computing system shown in Fig. 6.4.

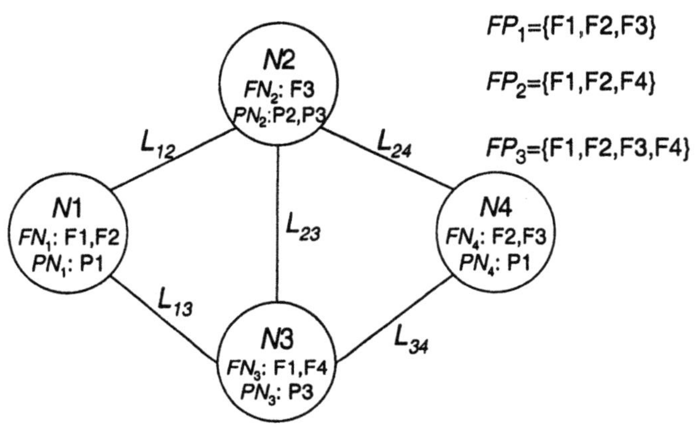

Fig. 6.4. A distributed computing system with four nodes and five links.

This distributed computing system consists of four processing nodes (N1,N2,N3,N4) that run three different programs (P1,P2,P3) distributed in redundant manner among the processing elements. Four data files (F1,F2,F3,F4) are also distributed in a redundant manner. FP_i (i=1,2,3) is the set of files that are required by the program P_i.

In Fig. 6.4, program P_1 can run successfully when either of N1 or N4 is working and it is possible to access the data files (F1,F2,F3). If P_1 is running on N1 which holds the files F1 and F2, it is required to access the file F3 which is resident at N2 or N4. That is, additional nodes and links are needed to have access to that required file (F3). Thus, the distributed program reliability depends on the reliability of all those involved processing nodes and communication links.

■

The distributed program reliability measures the reliability of one program in the system. However, for reliability of the distributed computing systems, it is important to obtain a global reliability measure that describes how reliable the system is for a given distribution of programs and files (Hariri & Mutlu, 1995). The definition of distributed system reliability is given below.

Definition 6.2. *Distributed system reliability* is the probability that all the distributed programs are executed successfully under the distributed computing environment.

As the distributed computing system depicted by Fig. 6.4, all three programs (P1,P2,P3) are required to be successfully achieved. Four data files (F1,F2,F3,F4) are needed when running those programs. Thus, the distributed system reliability here is the probability for all the three programs to be successfully executed meanwhile accessing to all the data files.

In order to estimate the distributed program/system reliability, some assumptions of the reliability model for the distributed computing system, see e.g. Kumar *et al.* (1986), are given below:

Assumptions:

1) Each node or link in the distributed computing system has two states: operational or faulty.
2) If a link is faulty, information cannot be transferred through it.
3) If a node is faulty, the program contained in the node cannot be

successfully executed, the files saved in it cannot be accessed by other nodes, and the information is not able to be transferred through it.

4) The probability for a processing node (N_i) to be operational is constant, which is denoted by p_i and $q_i = 1 - p_i$.

5) The probability for a communication link (L_{ij}) to be operational is also constant, which is denoted by p_{ij} and $q_{ij} = 1 - p_{ij}$.

6) Failures of all the nodes and links are statistically independent from each other.

It is indicated in Lin & Chen (1997) that computing distributed reliability is an *NP*-hard problem (Valiant, 1979) even when the distributed computing system is restricted to simple structures such as series-parallel, a tree, a star etc. Hence, general and effective analytical tools are required to evaluate its reliability.

6.2.2. Kumar's analytical tool

This analytical tool was presented by Kumar *et al.* (1986), which is based on Minimal File Spanning Tree (*MFST*). In general, the set of nodes and links involved in running the given program and accessing its required files form a tree. Such tree is called File Spanning Tree (*FST*) defined below.

Definition 6.3. *File Spanning Tree* is a spanning tree that connects the root node (the processing element that runs the program under consideration) to some other nodes such that its vertices hold all the required files for executing that program.

The smallest dominating file spanning tree is called *MFST* and its definition is given below.

Definition 6.4. A *Minima File Spanning Tree*, denoted by $MFST_i$, is an *FST* such that there exists no other file spanning tree, say FST_j, which is a subset of $MFST_i$.

An example of the *FSTs* and *MFSTs* is illustrated below.

Example 6.2. Continue considering the distributed system of Fig. 6.4. The following are some *FSTs* that make P_1 run successfully on N_1: $\{N_1, N_2, L_{12}\}$; $\{N_1, N_2, N_3, L_{12}, L_{23}\}$; $\{N_1, N_2, N_4, L_{12}, L_{24}\}$; $\{N_1, N_2, N_3, L_{13}, L_{23}\}$; $\{N_1, N_3, N_4, L_{13}, L_{34}\}$; $\{N_1, N_2, N_3, N_4, L_{12}, L_{23}, L_{34}\}$; $\{N_1, N_2, N_3, N_4, L_{12}, L_{24}, L_{34}\}$; $\{N_1, N_2, N_3, N_4, L_{13}, L_{23}, L_{24}\}$; $\{N_1, N_2, N_3, N_4, L_{13}, L_{34}, L_{24}\}$. Likewise, there will be several other *FSTs* when program P_1 runs on N_4.

The file spanning tree $\{N_1, N_2, N_4, L_{12}, L_{24}\}$ is not minimal because its subset $\{N_1, N_2, L_{12}\}$ is also an *FST*. We are interested in finding all the *MFSTs* to run a distributed program. For P_1 to run on either N_1 or N_4, four *MFSTs* are contained. They are

$$\{N_1, N_2, L_{12}\};\ \{N_1, N_2, N_3, L_{13}, L_{23}\};\ \{N_3, N_4, L_{34}\};\ \{N_2, N_3, N_4, L_{23}, L_{24}\}.$$

Anyone of these four *MFSTs* can provide a successful execution of the program under consideration when all elements are working.

■

From the above example, it can be seen that the distributed program can run successfully if any one of the *MFSTs* is operational. Hence, the distributed program reliability can be generally described in terms of the probability having at least one of the *MFSTs* operating as

DPR=Pr(at least one *MFST* of a given program is operational)

This can be written as

$$DPR = \Pr\left(\bigcup_{j=1}^{n_{mfst}} MFST_j\right) \tag{6.1}$$

where n_{mfst} is the total number of *MFSTs* that run the given program. The evaluation of the reliability of executing a program on a distributed system can be determined by the following two stages.

Stage 1. Find all the Minimal File Spanning Trees:

The purpose of this stage is to search all the *MFSTs* in which the roots are the processing elements that run a program, say P_i. The minimal file spanning trees are generated in nondecreasing order of their sizes, where the size is defined as the number of links in an *MFST*. At first, all *MFSTs* of size 0 are determined; this occurs when there exist some processing nodes that run P_i and have all the needed files (which is denoted by the set FP_i) for its execution. Then, all *MFSTs* of size 1 are determined; these trees have only one link which connects the root node to some other node, such that the root node and the other node have all the files in FP_i. This procedure is repeated for all possible sizes of *MFSTs* up to *n*-1, where *n* is the total number of nodes in the system. The detailed description of the algorithm to search all the *MFSTs* is given by Kumar *et al.* (1986).

Stage 2. Apply a terminal reliability algorithm to evaluate distributed program reliability:

Here we find the probability that at least one *MFST* is working which means that all the edges and vertices included in it are operational. Any terminal reliability evaluation algorithm based on path or cutest enumeration can be used to obtain the distributed program reliability of the program under consideration.

The distributed system reliability can be written as the probability of the intersection of the set of *MFSTs* of each program, which is

$$DSR = \Pr\left(\bigcap_{m=1}^{M} MFST(P_m)\right) \tag{6.2}$$

where $MFST(P_m)$ denotes the set of all the *MFSTs* associated with the program P_m.

Example 6.3. An example using the above analytical tool to estimate distributed program/system reliability is illustrated below.

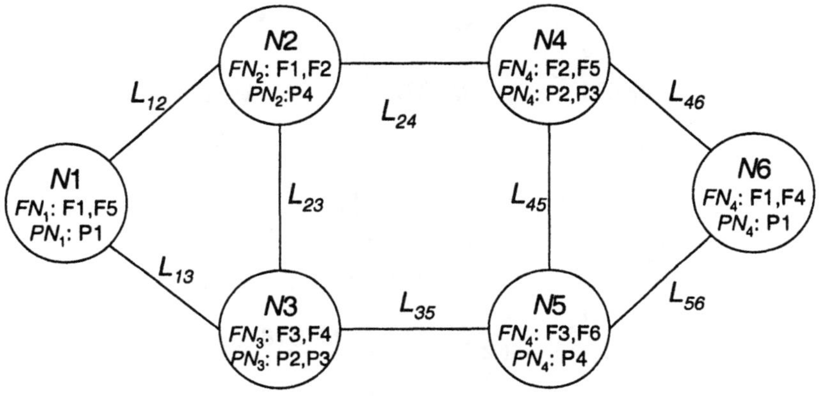

Fig. 6.5. A six-node distributed computing system.

The distributed computing system shown in Fig. 6.5 consists of six processing elements that can run four distributed programs and save six data files. The files needed for executing these programs are indicated in the following sets:

$$FP_1 = \{F_1, F_2, F_3\},\ FP_2 = \{F_2, F_4, F_6\},$$

$$FP_3 = \{F_1, F_3, F_5\},\ FP_4 = \{F_1, F_2, F_4, F_6\}.$$

Evaluating the distributed program reliability

We first derive the reliability of program P_1, denoted by $DPR(P_1)$. The program P_1 can run on either N1 or N6. Its *MFSTs* can be found by the step 1 as depicted by Fig. 6.6. The double-line circles represent the root node, the single-line circles represent the contained vertex, the number in the circle is the node number in the distributed computing system, and the files marked around the circles are reached new data files in that node.

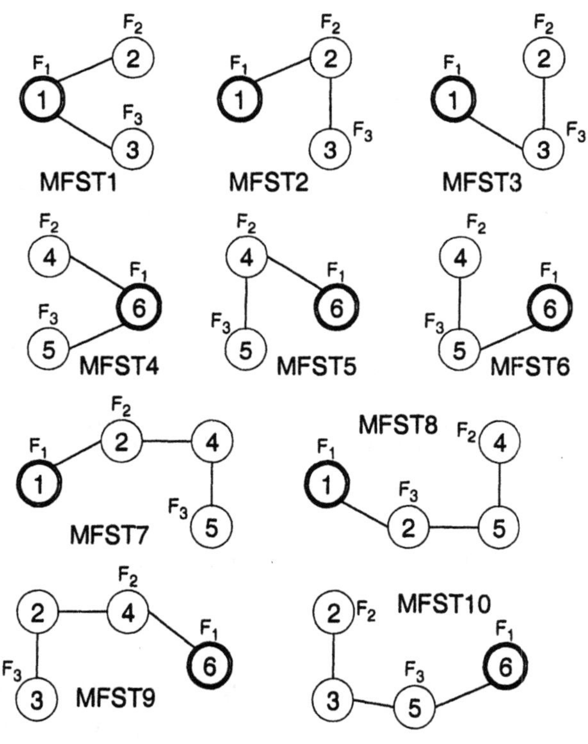

Fig. 6.6. ***MFSTs* for** $DPR(P_1)$.

Based on the generated *MFSTs* in Fig. 6.6, using the terminal reliability algorithm as step 2, we can obtain the reliability of program P_1. If we assume that all the elements (processing nodes and communication links), of the distributed computing system shown in Fig. 6.5 have the same reliability and equal to 0.9, then the reliability of executing P_1 is computed as 0.9378.

Evaluating the distributed system reliability

The first step in evaluating the distributed system reliability is to determine all the *MFSTs* for each program that can run on the system. The next step of this algorithm is to determine the set of all *MFSTs* that guarantee successful execution of all the programs by recursively intersecting the *MFSTs* of each program.

The final step is applying terminal reliability algorithm to obtain the following terms for the distributed system reliability. If we still assume that all the nodes and links have the same reliability, say 0.9, then the reliability of the distributed computing system is 0.842674.

■

6.2.3. A family of *FST* analytical tools

For analyzing the distributed program/system reliability, another family of File Spanning Tree (FST) analytical tools is further developed, which shows good efficiency for some specific problems.

The first *FST* analytical tool in this family was presented by Chen & Huang (1992) without considering node failures (i.e. just consider the communication failures of the network links). Hence, this analytical tool is only suitable for the distributed computing systems whose processing elements are perfect or highly reliable so that the probability for them to fail when working is negligible.

The main difference between this and above Kumar's tool is that Kumar's *MFST* starts from one root node and further expands the trees; and Chen-Huang's *FST* starts from the whole system graph and cuts links to prime the trees. Moreover, Kumar's tool requires to further use an additional terminal reliability algorithm to derive distributed program/system reliability but this *FST* tool can directly obtain the solution when priming the trees.

The basic concept of the *FST* analytical tool

The basic idea for the *FST* analytical tool is to find all disjoint *FSTs* in each size starting from the origin graph representing the distributed computing system. If we use an efficient method to cut one link each time from the graph at a different place to generate possible subgraphs recursively, then we are able to predict if each of these resulting subgraphs is an *FST* by examining the set of data files contained in this subgraph against the set of required data files for executing the distributed programs. This process can be repeated starting from graph size K, K-1,..., to 0 (where K is the number of links in the graph). Obviously, without an efficient method to remove appropriate links, the efficiency for the analytical tool could be very poor.

The method for cutting the graph plays an important role in finding the *FSTs* and in computing the reliability of the distributed computing system. The brief introduction for this method is given by the following five steps:

Step 1. Find a spanning tree from the current graph if necessary and compute ST_G (ST_G: a set of link states that can be used to construct the spanning tree of subgraph G), where each link L_{ij} has three states: 1) faulty state, denoted by 0; 2) operational state, denoted by 1; 3) undetermined state, denoted by *.

Step 2. Compute the vector LSP_G by $ST_G \wedge LS_G$, and convert vector LSP_G to the probability expression. (LSP_G: a set of link stats that can be used to compute the probability of subgraph G. The state condition could be either 1,0, or * as above; LS_G: a set of link states that represents the links' conditions in the current subgraph G).

Step 3. Cut the current graph according to the vectors ST_G and NC_G to obtain its subgraphs (or *FSTs*), where NC_G denotes a set of link states that indicate which link cannot be cut in subgraph G.

Step 4. Repeat steps 1 to 3 to compute each subgraph's vector LSP_G.

Step 5. The reliability of the distributed computing system graph is obtained by uniting all LSP_G vectors that are associated with all the *FSTs*.

Once the concept of finding all *FSTs* and computing the reliability of the distributed computing system is understood, the detailed algorithm for finding the *FSTs* and computing the reliability of the *FST* was illustrated in Chen & Huang (1992). An example for the *FST* analytical tool is illustrated below.

Example 6.4. Consider the simple distributed computing system in Fig. 6.4 again. We use the *FST* reliability analytical tool to analyze the distributed program/system reliability. For the program P_1, its reliability $DPR(P_1)$ is evaluated by the splitting snapshot of subgraphs generated by the above *FST* tool.

To compute the reliability, sum all the disjoint terms represented by vectors LSP_G, and then

$$DPR(P_1) = p_{12} + q_{12}p_{13}p_{23} + q_{12}q_{23}p_{34} + q_{12}q_{13}p_{23}q_{24}p_{34} + q_{12}q_{13}p_{23}p_{24}$$

Similarly, the distributed system reliability can be obtained from the above *FST* tool as

$$\begin{aligned} DSR = {} & p_{12}p_{23} + p_{12}p_{13}q_{23} + p_{12}p_{13}q_{23}p_{24}p_{34} + q_{12}p_{13}p_{23} \\ & + q_{12}q_{23}p_{24}p_{34} + q_{12}p_{13}p_{23}q_{24}p_{34} + q_{12}q_{23}p_{23}p_{24} \end{aligned} \tag{6.4}$$

If we assume all the links have the same reliability 0.9, then the

$$DPR(P_1)=0.99891 \text{ and } DSR=0.9963$$

■

The *FST-SPR* (Series and Parallel Reduction) analysis

How to speed the reliability evaluation process up is the major concern of the proposed analytical tool. The basic principle of speeding the reliability evaluation is to generate correct *FSTs* with less cutting steps. There are four methods presented by Chen & Huang (1992), which can be used interchangeably to speed the reliability evaluation. These methods are nodes merged, parallel reduction, series reduction, and degree-2 reduction as described below.

1) *Nodes merged* occurs when a probability subgraph has bit value 1 in its *LS* vector, i.e. the corresponding link must be operational in all its subgraphs. Hence the two nodes connected by this link can be merged into one node together with the link itself.

2) *Parallel reduction* occurs when a probability subgraph contains two or more links between two nodes. With connectivity property, these redundant links can be reduced to one link and the operational probability is replaced by

$$p_{ij} = 1 - \prod_k (1 - p_{ijk}) \tag{6.5}$$

where p_{ijk} is the operational probability for the k:th link between nodes i and j.

3) *Series reduction* occurs when a probability subgraph has a node, with node degree=2 (i.e. two links connect to this node), that contains no data file required for executing the distributed program. Since such a node, after deletion, still does not affect the correct *FST* generation, we can remove this node and reduce two links that connect to its neighboring nodes into one link. The new operation probability between the two neighboring nodes (i and j) is replaced by

$$p_{ij} = p_{ik} p_{kj} \tag{6.6}$$

where k is the deleted node.

4) *Degree-2 reduction* occurs when a probability subgraph has a node, with node degree = 2, that is not a leaf node of any *MFST* of the current graph. Since this node is not a leaf node of any *MFST*, then the two adjacent links of this node must be working or fail simultaneously, thus we can remove this node and reduce two links that connect to its neighboring nodes into one link, and copy the data files and programs in this node to either of its two neighboring nodes. The new operation probability between the two neighboring nodes (i and j) is replaced by

$$p_{ij} = p_{ik} p_{kj} \tag{6.7}$$

where k is the deleted node.

Note again, similar to *FST* analytical tool, this *FST-SPR* also assumes that the processing elements (i.e., nodes) in the distributed computing system is perfect. Hence, this analytical tool is also only suitable for the distributed computing systems whose processing elements are perfect or highly reliable so that the probability for them to fail is negligible when running the programs.

An example for the reduction methods of the *FST-SPR* is shown below.

Example 6.5. Suppose there is a subgraph generated as depicted by Fig. 6.9.

We need to compute the reliability of program P_1 which requires data files F1,F2,F3,F4 for completing its execution. The states of all the links are represented by different types of lines (dashed line: *failure*; double line: *operational*; single line: *undetermined*) and also by vectors *LS* and *NC*. The following are reduction steps for speedup the *FST* generation.

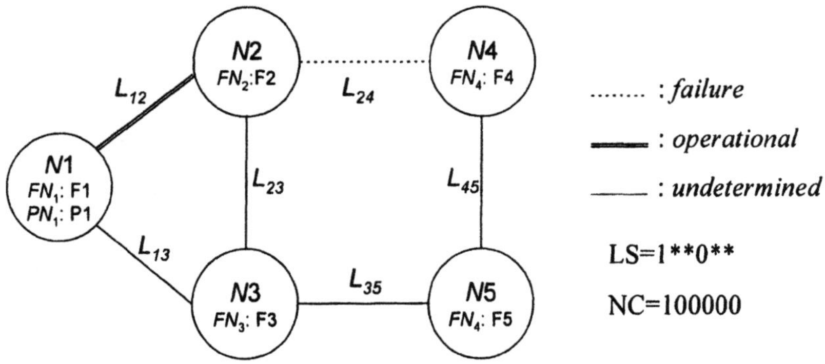

Fig. 6.7. A subgraph during reliability evaluation process.

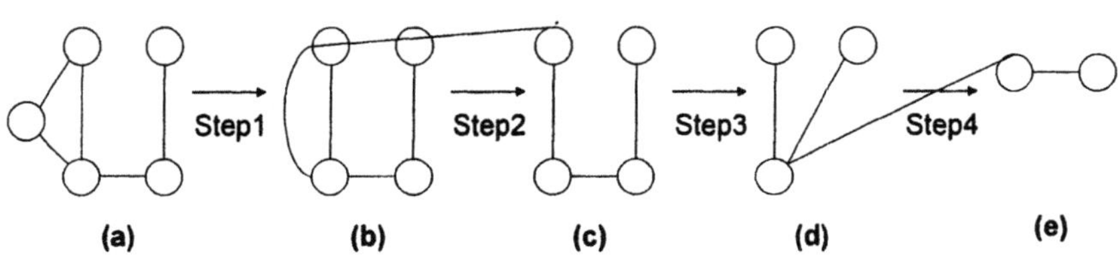

Fig. 6.8. Reduction steps for the subgraph of Fig. 6.7.

Step 1: Since link L_{12} can no longer be cut and must be up for the rest of its subgraph generation due to LS=1**0**, we apply nodes merged reduction on nodes N1 and N2. The resulting subgraph(b) is shown in Fig. 6.8(b).

Step 2: A parallel reduction can be applied on the resulting subgraph (from step 1) since links L_{13} and L_{23} are parallel. The new resulting subgraph (c) is shown in Fig. 6.8(c).

Step 3: A series reduction occurs since node N5 contains no data files for the execution of P_1. The new resulting subgraph (d) is shown in Fig. 6.8(d).

Step 4: A degree-2 reduction occurs since node N3 is not a leaf node of any *MFST*. The final subgraph (e) after these reductions is also shown in Fig. 6.8(e).

■

6.3. Homogeneously Distributed Software/Hardware Systems

A typical kind of application on distributed systems has a homogeneously distributed software/hardware structure. The physical system is assumed to contain *N* software subsystems (SW1-SW*N*) running on *N* hosts (HW1-HW*N*) as depicted in Fig. 6.9.

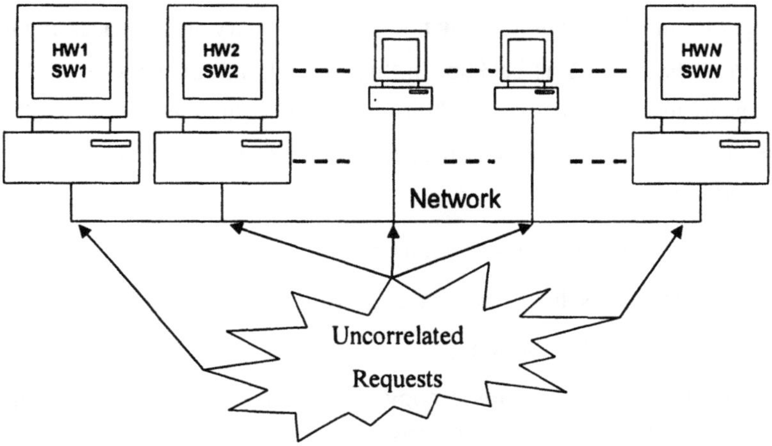

Fig. 6.9. A general homogeneous distributed software/hardware system.

That is, identical copies of distributed application software run on the same type of hosts, called Homogeneous Distributed Software/Hardware System. This system may be implemented to provide services for uncorrelated random requests of customers.

In this system, the software is usually improved. Since the system considers combined software and hardware failures as well as maintenance process, its reliability cannot be simply estimated by the above analytical tools for computing the distributed program reliability. The availability models and analyses of the homogeneous distributed software/hardware system are studied here.

6.3.1. Availability model

Actually, homogeneous distributed software/hardware system is a type of cluster system, which is a collection of computers in which any member of the cluster is capable of supporting the processing functions of any other member Mendiratta (1998) and Lyu & Mendiratta (1999). A cluster has a redundant $n+k$ configuration, where n processing nodes are necessary and k processing nodes are in spare state, serving as backup. In this subsection, our model is a cluster of N homogeneous hosts that are working in parallel. This means that if all of the N hosts failed, the system fails. Otherwise whenever 1 host can work, the system is still working.

The following are the assumptions concerning this system:

(a) All the hosts have the same hardware failure rate λ_h arising from an exponential distribution.

(b) Each of the hosts runs a copy of the same software with a failure rate function $\lambda_s(t)$ of a given software model.

(c) Both the software and hardware have only two states, up (working state) and down (malfunctioning state), which means all the failures of software or hardware are crash failures.

(d) There are maintenance personnel to repair the system upon software or hardware malfunction. The repair time has an exponential distribution with parameter μ_s for software and parameter μ_h for hardware, respectively.

(e) All the failures involved (either software or hardware) are mutually independent.

(f) No two or more failures (either software or hardware) occur at the same time.

There are some real cases of homogeneously distributed software/hardware system in which all the hosts can work independently for random/unknown request. Such applications can be found in search engine system, telephone switching system and banking system, and so on. Most homogeneous distributed software/hardware systems that work independently under the case of uncorrelated random requests can implement our models.

Systems in practice can be complex and usually we have a multi-host situation. Lai *et al.* (2002) used a Markov process to model this type of system. Fig. 6.10 illustrates a partial system state transition of the Markov process, in which (i, j) is the state when i hosts suffer hardware failures and j hosts suffer software failures.

The corresponding Chapman-Kolmogorov differential equation for the probability that the system is in the state (i,j) at time t is, for $i, j \neq 0, N; i + j \leq N-1$,

$$
\begin{aligned}
P'_{i,j}(t) = {} & \mu_h P_{i+1,j}(t) + (N-i-j+1)\lambda_h P_{i-1,j}(t) \\
& + (N-i-j+1)\lambda_s(t) P_{i,j-1}(t) + \mu_s P_{i,j+1}(t) - x_{i,j} P_{i,j}(t)
\end{aligned}
\tag{6.8}
$$

where

$$
x_{i,j} = \mu_s + (N-i-j)\lambda_h + (N-i-j)\lambda_s(t) + \mu_h \tag{6.9}
$$

The initial conditions are

$$
P_{0,0}(0) = 1 \text{ and } P_{i,j}(0) = 0, \text{ for } i, j \neq 0 \tag{6.10}
$$

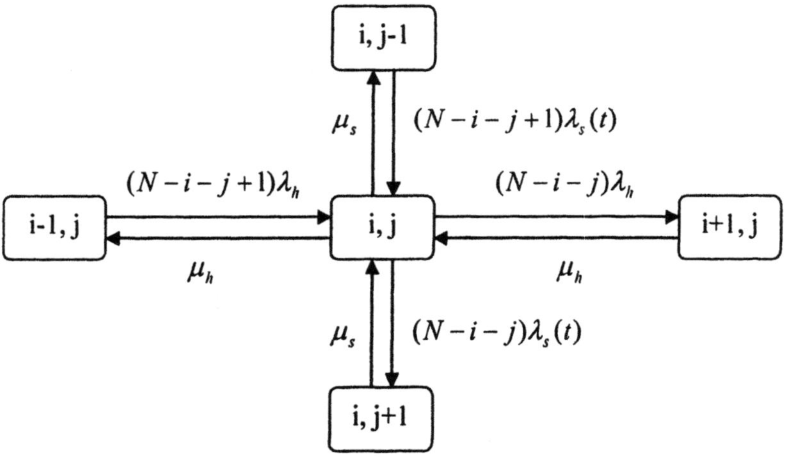

Fig. 6.10. The partial state transition graph for the *N*-host system.

The boundary conditions are:

$$
\begin{aligned}
&P'_{0,0}(t) = \mu_h P_{1,0}(t) + \mu_s P_{0,1}(t) - N[\lambda_s(t) + \lambda_h]P_{0,0}(t) \\
&P'_{0,j}(t) = \mu_h P_{1,j}(t) + (N-j+1)\lambda_s(t)P_{0,j-1}(t) + \mu_s P_{0,j+1}(t) \\
&\qquad -\{\mu_h + \mu_s + (N-j)[\lambda_s(t) + \lambda_h]\}P_{0,j}(t) \text{ for } j = 1,2,\ldots,N-1 \\
&P'_{i,0}(t) = \mu_h P_{i+1,0}(t) + (N-i+1)\lambda_h P_{i-1,0}(t) + \mu_s P_{i,1}(t) \\
&\qquad -\{\mu_h + \mu_s + (N-i)[\lambda_s(t) + \lambda_h]\}P_{i,0}(t) \text{ for } i = 1,2,\ldots,N-1 \\
&P'_{i,j}(t) = (N-i-j+1)\lambda_h P_{i-1,j}(t) + (N-i-j+1)\lambda_s(t)P_{i,j-1}(t) \\
&\qquad -(\mu_s + \mu_h)P_{i,j}(t) \qquad \text{for } i+j = n+k; 0 < i,j < n+k \\
&P'_{N,0}(t) = \lambda_h P_{N-1,0}(t) - \mu_h P_{N,0}(t) \\
&P'_{0,N}(t) = \lambda_s(t)P_{0,N-1}(t) - \mu_s P_{0,N}(t)
\end{aligned}
\tag{6.11}
$$

The system availability for the *N*-host based system can be calculated by

$$A(t) = \sum_{i+j<N} P_{i,j}(t) \tag{6.12}$$

Here, we assume each copy of software suffers a failure rate of the JM model (Jelinski & Moranda, 1972), i.e.,

$$\lambda_s(t) = k_t \phi \tag{6.13}$$

To solve the above differential equations, we need to know the expected number of remaining software faults (k_t). However, since k_t changes with software debugging, it is usually a function of time. We have used the following scheme for the numerical calculation, as shown by Lai *et al.* (2002). According to the JM model, the probability of software having *k* remaining faults at time *t* is

$$P(k,t) = \binom{K_0}{k} \exp(-k\phi t) \cdot [1 - \exp(-\phi t)]^{K_0 - k} \quad \text{for } 0 \le k \le K_0 \tag{6.14}$$

Based on this equation, the expected number of remaining software faults at time *t* can be computed as

$$k_t = \sum_{k=0}^{K_0} k \cdot P(k,t)$$

The system availability can be computed using any available numerical algorithm to solve the differential equations. An example using our above Markov model to analyze availability of homogeneous distributed software/hardware system is numerically illustrated below.

Example 6.6. We assume that the hardware failure rate is 0.02 and software failure rate per fault is 0.006. The repair rate for hardware is 0.1 while that for software is 0.12. Fig. 6.11 depicts the result of system availability of a triple-host system with different number of initial faults.

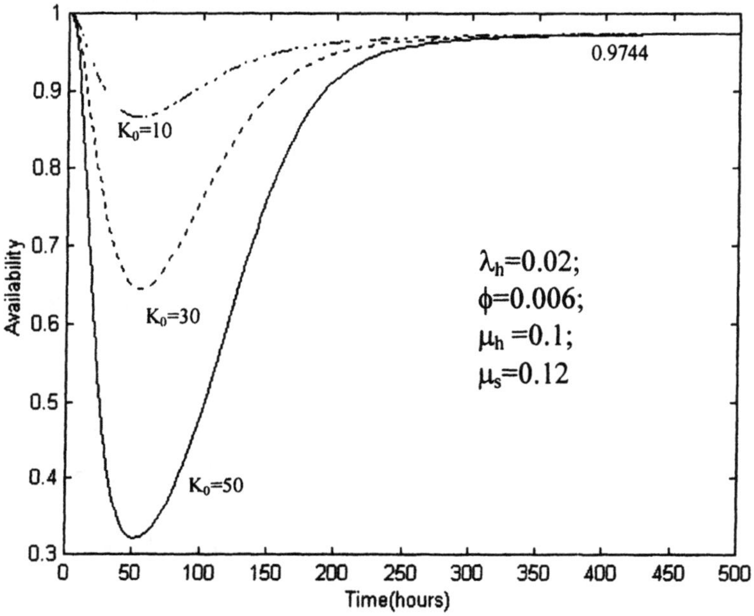

Fig. 6.11. A typical curve of the system availability function.

It can be seen from Fig. 6.11 that the system availability reaches the lowest point at an early stage. This is because a large number of faults are identified when software system testing begins. System availability starts recovering after the lowest point and approaches a certain value less than 1 asymptotically after a longer period of time. This is because identified faults are fixed and as a result software failure rate decreases.

■

6.3.2. Model of the imperfect debugging process

In the above section, the homogeneous distributed software/hardware system model assumed that the debugging process was a perfect one. It is possible in reality that the fault that is supposed to have been removed may cause a failure again. It may be due to the spawning of a new fault because of imperfect debugging, see e.g. Fakhre-Zakeri & Slud (1995), Sridharan & Jayashree (1998), Pham *et al.* (1999) and Tokuno & Yamada (2000).

Markov modeling

The assumptions used in this imperfect debugging model are almost the same as the assumptions (*a-f*) in earlier model except that the following assumption is added.

(g) When a software failure occurs, instantaneously repair starts with the following debugging probabilities:

The software fault content is reduced by one with probability p.

The software fault content remains unchanged with probability r.

The software fault content is increased by one with probability q.

This assumption is same as the birth-death process that was introduced in Kremer (1983).

Fig. 6.12 illustrates a partial system state transition, in which (i, j, k) is the state when i hosts suffer hardware failures, j hosts suffer software failures and k is the number of remaining software faults at that time. Here N is the total number of hosts in the system.

The corresponding Chapman-Kolmogorov differential equation for the probability that the system is in the state (i, j, k), $P_{i,j,k}(t)$, at time t can be obtained. They can be solved numerically or analytically in some cases. The system availability at time t can be calculated as

$$A(t) = \sum_{i=0}^{N-1} \sum_{j=0}^{N-i-1} \sum_{k=0}^{K_0} P_{i,j,k}(t) \tag{6.15}$$

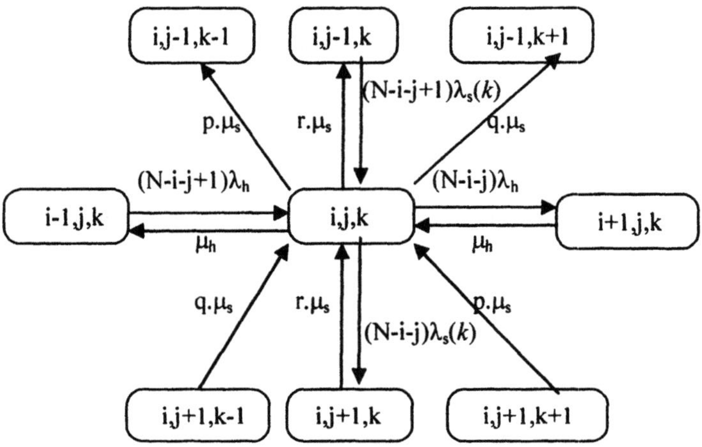

Fig. 6.12. The state transition graph for the *N*-host system.

Although those differential equations can be solved, the procedure becomes difficult when the number of hosts is large. Hence, some computing tools can be used to solve them. An example is illustrated below.

Example 6.7. In this numerical example, the software failures are assumed to follow the JM-model. For the multi-host systems with different number of hosts, the system availability functions can be obtained numerically. The curves of system availability functions for (N=2,3,4,5) are depicted in Fig. 6.13 with parameters

$$\mu_h=0.1536,\ \mu_s=0.1331,\ p=0.831,\ q=0.078,$$
$$r=0.091,\ K_0=42,\ \phi=0.0013 \text{ and } \lambda_h=0.005$$

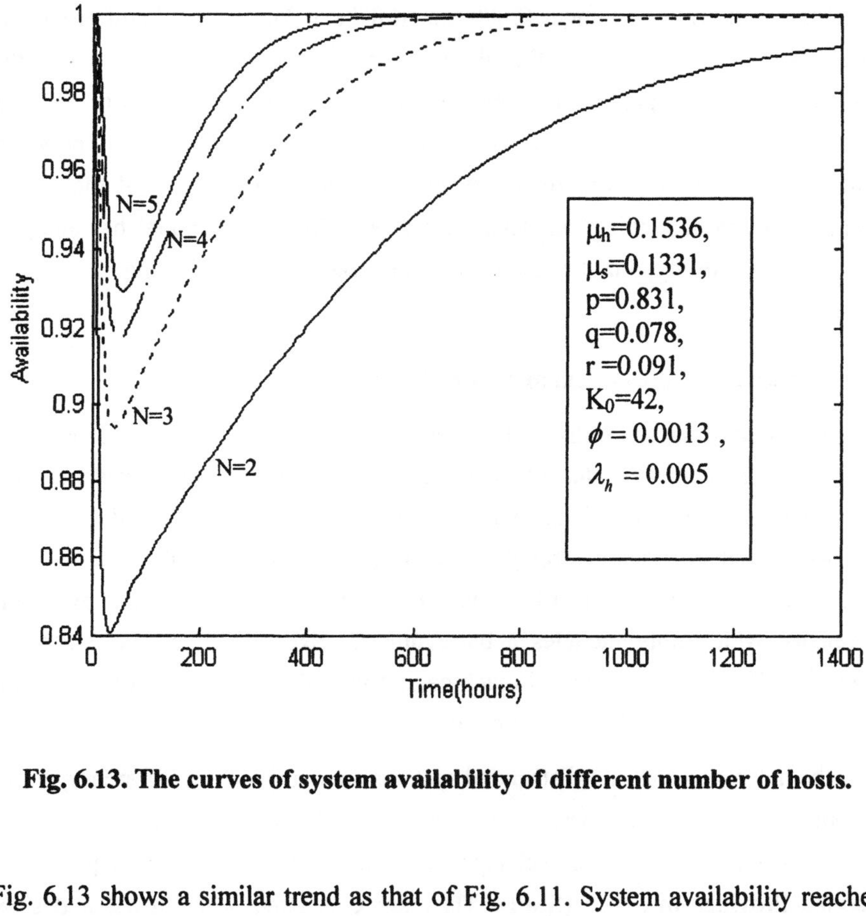

Fig. 6.13. The curves of system availability of different number of hosts.

Fig. 6.13 shows a similar trend as that of Fig. 6.11. System availability reaches the lowest point at an early stage. After that period, system availability starts recovering because identified faults are fixed and as a result software failure occurrence rate decreases.

6.4. Centralized Heterogeneous Distributed Systems

Most of the distributed service systems can be modeled by a centralized heterogeneous distributed system. This type of distributed systems consists of heterogeneous subsystems that are managed by a control center, see e.g. Hussain & Hussain (1992) and Langer (2000, pp. 188-217). The system is different from the systems and models in the above sections, because those models either assumed constant operational probability without reliability growth or excluded the network reliability. However, the system incorporates not only the hardware/software/network reliability but also the improvement of the control center through debugging/maintenance process. Dai *et al.* (2003a) has analyzed its service reliability, and the results are summarized in the following.

6.4.1. Service of the system and its reliability

The structure of the Centralized Heterogeneous Distributed System is depicted by Fig. 6.14. The control center may consist of many servers. These servers support a virtual machine. The virtual machine can manage programs and data from heterogeneous subsystems through virtual nodes. The virtual nodes can mask the differences among various platforms. They are a kind of virtual executing elements, which only includes a basic unit for executing data, i.e. CPU and Memory. The entities of virtual machine and virtual nodes are supported by the software and hardware in the control center.

The heterogeneous sub-distributed systems are composed of different types of computers with various operating systems connected by different topologies of networks. These subsystems exchange data with virtual machine through System Service Provider Interface (SSPI). They are connected with virtual nodes by routers. They can cooperate to achieve a distributed service under the management of the virtual machine.

In fact, most of service systems can be categorized as centralized heterogeneous distributed systems such as the example of military system shown in Fig. 6.2.

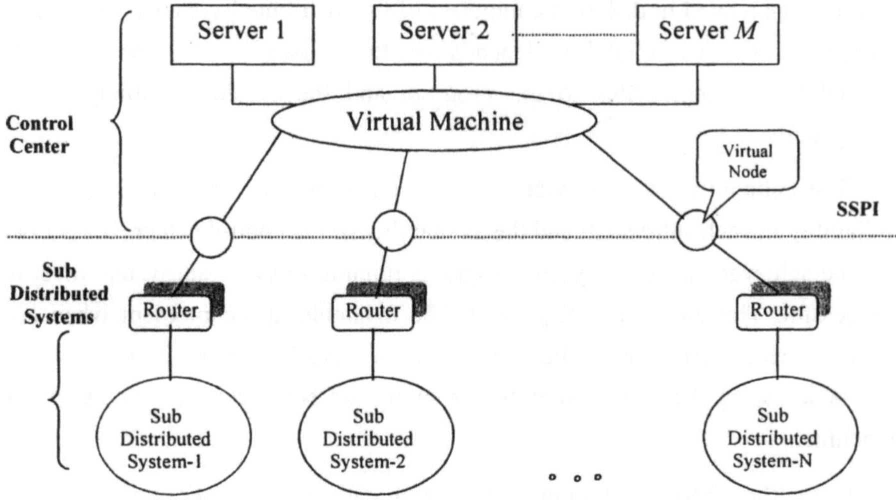

Fig. 6.14. Structure of the centralized heterogeneous distributed service system.

The whole process for a service in a distributed system is repeated so the reliability analysis of a distributed service is crucial for a distributed system. The distributed service reliability is defined as below.

Definition 6.5. *Distributed service reliability* is the probability for a service to be successfully achieved in a distributed computing system.

6.4.2. Model of distributed service reliability

In a distributed service system, a service includes various distributed programs completed by diverse computers. Some later programs might require several precedent programs to be completed. Every program requires a certain execution time. The execution of some programs might require certain input files that are saved or generated in different computers of the distributed systems. The overall distributed service reliability depends on the reliability of a program, the availability of input files to the program and the system reliability of the subsystem.

The reliability of a service is determined by the distributed programs reliability in each subsystem and the availability of the control center. If a service can be achieved successfully, the programs running in every subsystem must be successful. The virtual machine should be available at the moment when any program needs certain input file prepared in the virtual machine. It has to be also available during the period when the programs are being executed in the virtual machine.

It can be obtained through the critical path method, see e.g. Hillier & Lieberman (1995), that the time point when the programs require the files prepared in the virtual machine (T_{bf}^{j}), j=1,2,...J. We can also obtain the starting time when the programs run in the virtual machine (T_{bp}^{k}) and the corresponding execution time period for those programs (T_{ex}^{k}), k=1,2,...,K.

It is noted that $A(t)$ is the availability of the virtual machine at time t. We also assume that the programs require input files at the beginning time, T_{bf}^{j}, so the availability of the input files can be calculated as

$$P_f(j) = A(T_{bf}^{j}),\ j=1,2,...J \tag{6.16}$$

It is assumed that the virtual machine has to be available from the beginning to the end when a program runs in it; otherwise, the program fails. The average availability of the programs, which start at time T_{bp}^{k} with the execution time period T_{ex}^{k}, can be calculated as

$$P_{pr}(k) = \int_{T_{bp}^{k}}^{T_{bp}^{k}+T_{ex}^{k}} A(t)dt \,/\, T_{ex}^{k},\ k=1,2,...,K \tag{6.17}$$

Let N be the number of subsystems. The distributed system reliability for the i:th subsystem is denoted by DSR_i (i=1,2,…,N) where the virtual machine is viewed as a perfect node in each sub-distributed systems at first. The DSR_i (i=1,2,…,N) can be computed by the various algorithms presented in the previous section. Then, the availability of the virtual machine is incorporated into the distributed service reliability together with the DSR_i.

In order to calculate distributed service reliability, some additional assumptions on statistical independence are needed:

1) DSR_i (i=1,2,…,N) is assumed to be mutually independent.

2) The files prepared in the virtual machine are also mutually independent.

3) The programs running in the virtual machine are mutually independent.

Although the independence assumption may not always be true, they are first order approximation.

The distributed service reliability function to the initial time, t_b, can be calculated by

$$R_s(t_b) = \prod_{i=1}^{N} DSR_i \prod_{j=1}^{J} P_f(j) \prod_{k=1}^{K} P_{pr}(k) \tag{6.18}$$

Eq. (6.18) can be explained as follows. The virtual machine can be viewed as a perfect node in calculating DSR_i without considering the availability of prepared files and executed programs in it. Thus, the service reliability is the whole distributed system reliability $\prod_{i=1}^{N} DSR_i$ multiplied by the availability of files and programs in the virtual machine.

Furthermore, the availability of files and programs in the virtual machine can be expressed as the product of $\prod_{j=1}^{J} P_f(j)$ and $\prod_{k=1}^{K} P_{pr}(k)$. Hence, the overall distributed service reliability function which is the product of all three quantities can be expressed as in the above equation.

6.4.3. Algorithm for distributed service reliability

In applying the general approach, we will need the system structure and then the above model can be used. The algorithm for the calculation of the distributed service reliability can be presented as the following six steps:

Step 1: Identify the structure of Centralized Heterogeneous Distributed System and relationship between programs and files.

Step 2: Obtain the availability function of the virtual machine with any existing models.

Step 3: Let the virtual machine to be a perfect node in every subsystem and calculate DSR_i (i=1,2,...,N).

Step 4: Using the critical path method to determine T_{bf}^j (j=1,2,...J) and T_{bp}^k, T_{ex}^k (k=1,2,...,K).

Step 5: Calculate $P_f(j)$ and $P_{pr}(k)$.

Step 6: Calculate the distributed service reliability function at time t_b.

Note that we can implement different models and methods to calculate distributed service reliability. For subsystems, the DSR_i can be calculated through the algorithms, e.g. *MFST* (Kumar *et al.*, 1986), *FST* (Chen & Huang, 1992), *HRFST* (Chen *et al.*, 1997), etc. For the availability function of the virtual machine $A(t)$, it can be calculated through the models presented by Lai *et al.* (2002).

6.5. Notes and References

In the distributed computing systems, the group of *MFST* algorithms is further developed. Kumar (1988) proposed a "Fast Algorithm for Reliability Evaluation" that used a connection matrix to represent each *MFST* and proposed some simplified techniques for speeding up the analysis process. Then, Kumar & Agrawal (1996) further introduced "Distributed Program/System Performance

Index" which can be used to compare networks with different features for application execution.

For the group of *FST* analytical tools, Lin *et al.* (1999a) further presented an efficient algorithm for reliability analysis of distributed computing systems. The efficient algorithm was studied specifically in the different kinds of network topologies such as star topologies (Chang *et al.*, 2000 and Lin, 2003) and ring-type topologies (Lin *et al.*, 2001). This group of analytical tools can be further extended to allow the failures of imperfect nodes, see e.g. Ke & Wang (1997) and Lin *et al.* (1999b).

Other than the above two groups of analytical tools, Lopez-Benitez (1994) also presented a modeling approach based on stochastic Petri nets to estimate the reliability and availability of programs in a distributed computing system. Later, Chen *et al.* (1998) presented a Markov model to study the distributed system reliability with the information on time constraints. Malluhi & Johnston (1998) developed a distributed parallel storage system to achieve scalability and high data throughput. Fricks *et al.* (1999) proposed an analytic approach, based on the Markov regenerative processes and the Petri nets, to compute the response-time distribution of operator consoles in a distributed process control environment. Das & Woodside (2001) evaluated the layered distributed software systems with fault-tolerant features. Yeh & Chiu (2001) proposed a reversing traversal method to derive a *k*-node distributed system under capacity constraint. Chiu *et al.* (2002) recently developed a reliability-oriented task allocation scheme for the distributed computing systems. Mahmood (2001) discussed the task allocation algorithms for maximizing reliability of heterogeneous distributed computing systems.

Fahmy (2001) considered reliability evaluation in distributed computing environments by using the concept of Analytical Hierarchy Process (AHP). Lanus *et al.* (2003) presented hierarchical composition and aggregation models based on Markov reward models to study the state-based availability and performability of distributed systems. Yeh (2003) extended the distributed system reliability by introducing a multi-state concept.

CHAPTER 7

RELIABILITY OF GRID COMPUTING SYSTEMS

Grid computing is a recently developed technique for complex systems with large-scale resource sharing, wide-area program communicating, and multi-institutional organization collaborating etc. Many experts believe that the grid technologies will offer a second chance to fulfill the promises of the Internet (Forster *et al.*, 2002). However, it is difficult to analyze the grid reliability due to its highly heterogeneous and wide-area distributed characteristics.

This chapter first presents a brief discussion of the Grid computing system. A general grid reliability model is then constructed. We also present approaches to compute the grid reliability by incorporating various aspects of the grid structure including the resource management system, the network and the integrated software/resources.

7.1. Introduction of the Grid Computing System

7.1.1. Grid technology

The term "Grid" was created in the mid 1990s to denote a proposed distributed computing infrastructure for advanced science and engineering (Foster & Kesselman, 1998). Grid concepts and technologies were first developed to enable resource sharing within far-flung scientific collaborations. Applications include collaborative visualization of large scientific datasets (pooling of expertise), distributed computing for computationally demanding data analyses (pooling of compute power and storage), and coupling of scientific instruments with remote computers and archives (increasing functionality as well as availability).

The real and specific problem that underlies the Grid concept is coordinated resource sharing and problem solving in dynamic, multi-institutional virtual organizations (Foster *et al.*, 2001). The sharing that we are concerned with is not primarily file exchange but rather direct access to computers, software, data, and other resources. This is required by a range of collaborative problem-solving and resource-brokering strategies emerging in industry, science, and engineering. This sharing is highly controlled, with resource providers and consumers defining what is shared, who is allowed to share, and the conditions under which the sharing occurs. A set of individuals or institutions are defined by such sharing rules form what is usually called *virtual organization* (VO).

For example, in a data grid project thousands of physicists at hundreds of laboratories could be involved. They can be divided into different virtual organizations according to their locations or functions. It is depicted by Fig. 7.1.

In this case, virtual organizations can vary tremendously in their purpose, scope, size, duration, structure, community, and sociology. A careful study of underlying technology requirements, however, leads us to identify a broad set of common concerns and requirements and current distributed computing technologies do not address the concerns and requirements of the grid.

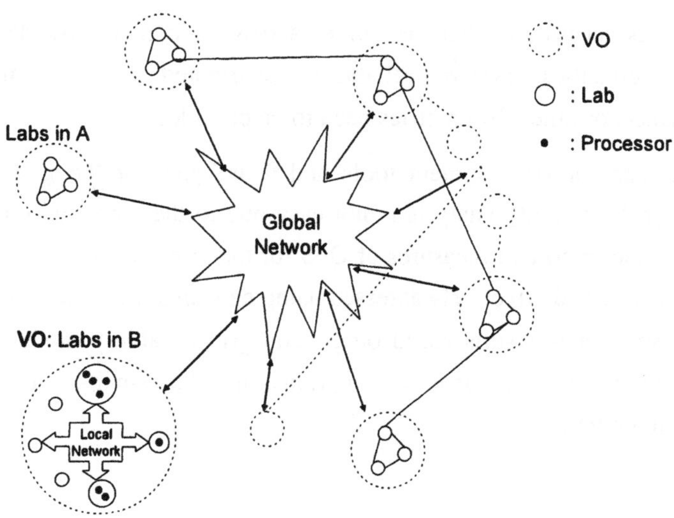

Fig. 7.1. A grid computing system containing many virtual organizations.

Over the past several years, research and development efforts within the grid community have produced protocols, services, and tools that address precisely the challenges that arise when we seek to build scalable virtual organizations, e.g. Foster & Kesselman (1998), Foster *et al.* (2001, 2002), Frey *et al.* (2002) and Buyya *et al.* (2003).

Because of their focus on dynamic, cross-organizational sharing, Grid technologies complement rather than compete with the existing distributed computing technologies. For example, enterprise distributed computing systems can use the grid technologies to achieve resource sharing across institutional boundaries. The grid technologies can also be used to establish dynamic markets for computing and storage resources.

The continuing decentralization and distribution of software, hardware, and human resources make it essential that we achieve the desired quality of service (QoS) on resources assembled dynamically from enterprise, service provider, and customer systems. This also requires new abstractions and concepts that let applications access and share resources across wide-area networks. Common security semantics, system reliability, distributed resource management performance, or other QoS metrics need to be provided.

Although the development tools and techniques for the grid have been studied, grid reliability analysis is not easy due to the complexity of the grid. As one of the important measures of QoS for the grid, the grid reliability needs to be precisely and effectively assessed using new analytical tools. This chapter presents some new results based on general grid reliability models that relax some unsuitable traditional assumptions in the small-scale distributed computing systems.

7.1.2. General architecture of grid computing system

The general architecture of the grid computing systems can be depicted as Fig. 7.2. The virtual node is a general unit in the grid, which can execute programs or share resources. Virtual nodes are connected with each other through the virtual links. Virtual organizations are made up of a number of virtual nodes.

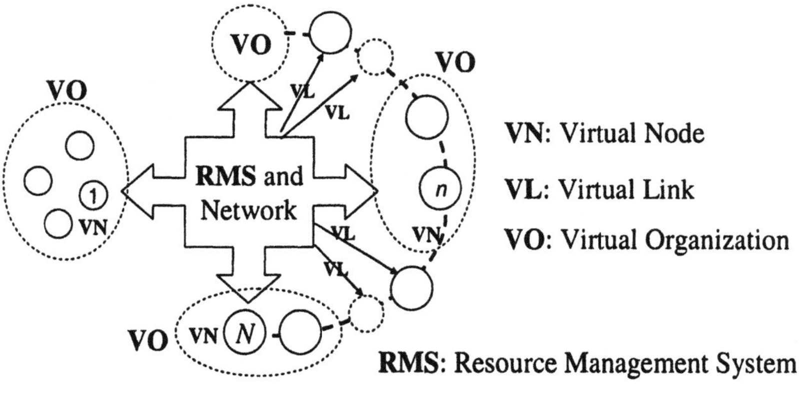

Fig. 7.2. General architecture of grid computing systems.

A grid system is designed to complete a set of programs/applications, so that to complete certain tasks. Executing those programs need use some resources in the grid. These programs and resources are distributed on the virtual nodes as in Fig. 7.2. A virtual link between two virtual nodes (*i* and *j*), denoted by *L*(*i*, *j*), is defined as a direct communication channel between the two nodes *i* and *j* without passing through other virtual nodes.

Let $\vec{U}_n$ represent the set of resources shared by the *n*:th virtual node and $\vec{V}_n$ represent the set of programs executed by the *n*:th virtual node, (*n*=1,2,...,*N*). We also assume that *M* programs denoted by $P_1, P_2, ..., P_M$ are running in the grid system. The required processing time for each program is denoted by $t(1), t(2), ..., t(M)$, respectively. The programs may use some necessary resources during their execution, which is in fact to exchange information between them through the network. These resources are denoted by $R_1, R_2, ..., R_H$ which is registered in a resource management system of the grid.

When a program requests certain remote resources, the resource management system receives these requests and matches the registered resources to the requests. It then instructs the program the sites of those matched resources. After the programs know the sites of their required resources, they begin to access to them through the network.

In an early stage, the grid reliability is mainly determined by the reliability of the resource management system, while in a later stage, the grid reliability is mostly affected by the reliability of the network for communicating or processing. The grid reliability model related to the two stages will be studied respectively in the following two sections. Then, Section 7.4 further integrates other components such as software and resources etc into the grid reliability analysis.

7.2. Grid Reliability of the Resource Management System

Before the programs begin to access to their required resources in the grid, they have to know the sites of those resources, which is managed by the resource management system. The resource management system of the grid, see e.g. Livny & Raman (1998), is to receive the resource requests from application programs, and then to match the requests with the registered resources.

7.2.1. Introduction of resource management system

For grid computing, the resource management system that manages its pool of shared resources is very important. This is especially the case for Open Grid Service Architecture, see e.g. Foster *et al.* (2002), that allows individual virtual organizations to aggregate their own resources on the grid.

The resource management system provides resource management services, which can be divided into four general layers as depicted by Fig. 7.3. They are program layer (A), request layer (B), management layer (C) and resource layer (D).

A. *Program layer*: The program layer represents the programs (or tasks) of the customer's applications. The programs describe their required resources and constraint requirements (such as deadline, budget, function etc).

B. *Request layer*: The request layer represents the program's requirement for the resources. This layer provides the abstraction of "program requirements" as a queue of resource requests.

C. *Management layer*: The management layer may be thought of as the global resource allocation layer and its principal function is to match the resource requests and resource offers so that the constraints of both are satisfied.

D. *Resource layer*: The resource layer represents the registered resources from different sites including the requirements and conditions.

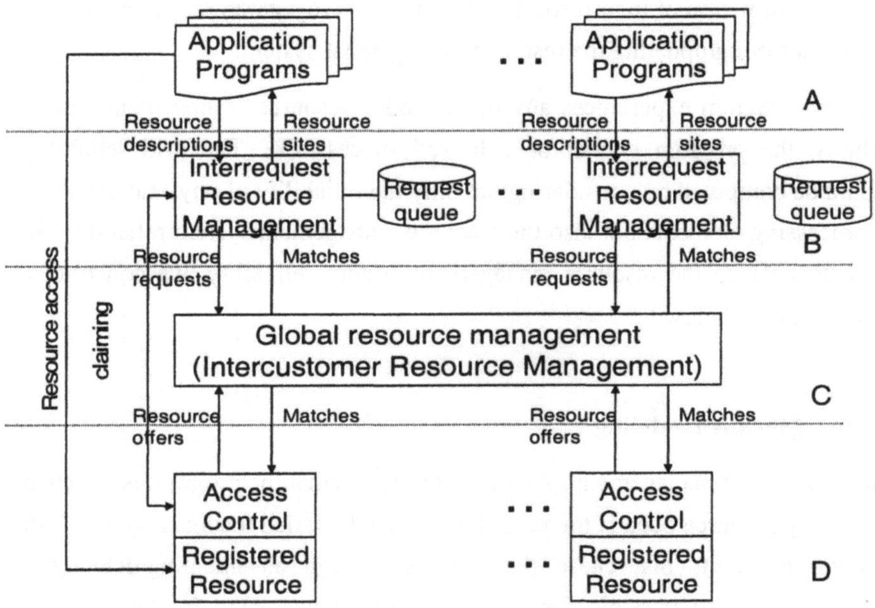

Fig. 7.3. Layers of resource management system.

In grid computing, failures may occur at any of the layers in the resource management system. For example,

1) In the program layer, the resource described by the program may be unclear or translated into wrong resource requests.

2) In the request layer, the request queue may be too long to be waited by the program (generating so called time-out failures), or some requests may be lost due to certain management faults.

3) In the management layer, the request may be matched to a wrong resource because of misunderstanding or faulty matching.

4) In the resource layer, the virtual organization may register wrong information of their resources or remove its registered resources without notifying/updating the resource management system.

If a grid program experiences any of the above resource management system failures, the program cannot be achieved successfully. The grid reliability should be computed by considering not only the reliability of physical networks or processing elements but also the resource management system reliability. In order to analyze the resource management system reliability, we construct a Markov model below.

7.2.2. Markov modeling

For the resource management system, if any failure that the program is matched to a wrong resource occurs, the program will send a failed feedback to it. It will remove the faults that cause the failures through an updating/debugging process. It is also possible for new faults to be generated in the resource management system such as some virtual organizations register wrong resources to it, etc. The assumptions for our resource management system reliability model are listed as follows:

1) The failures of resource management system follow an exponential distribution with parameter $\lambda(k)$ where k is the number of contained faults.

2) If any failure occurs, a fault that causes this failure is assumed to be removed immediately by an updating/debugging process, i.e. the time for removing the detected fault is not counted.

3) The resource management system may generate a new fault, and the occurrence of such event follows an exponential distribution with a constant rate ν.

According to the above assumptions, the reliability model of resource management system can be constructed by a continuous time Markov chain (CTMC). This Markov model depicted in Fig. 7.4 is a typical birth-death Markov process with infinite number of states, where state k represents k faults contained in the resource management system.

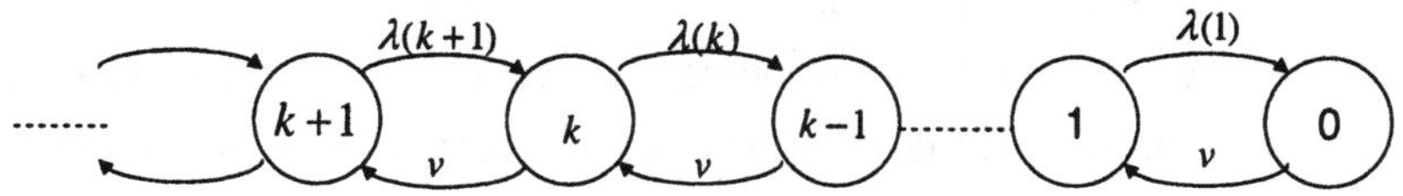

Fig. 7.4. CTMC for resource management system reliability model.

In this model, $\lambda(k)$ can be a function of the number of remaining faults k. Usually, $\lambda(k)$ is an increasing function to the number of remaining faults k. It is desired for a resource management system to be in service for a long time, especially for the Open Grid Service Architecture (Foster *et al.*, 2002), so the birth-death process of failures can be viewed as a long-run Markov process (Trivedi, 1982). After running for a long time, the expected death rate $\lambda(k)$ will approach to a steady value. The failure rate $\lambda(k)$ can be approximately viewed as a constant during a small enough time. An example is illustrated below.

Example 7.1. Consider a grid program denoted by P1 need access to remote resources. The time for resource management system to deal with its request is supposed t=15 seconds and the failure rate of resource management system at that time slot $\lambda = 0.0005$ per second. The reliability for the resource management system to deal with the request is computed as

$$R_{RMS}(P1) = \exp\{-\lambda \cdot t\} = 0.992528$$

Based on the long-run birth-death Markov process, this approximation of constant failure rate indicates a way to reasonably and dynamically update the failure rate at different time slots. The resource management system can count the number of failures, say n, reported by the grid programs between a relatively small time interval, say Δt, and dynamically updates the value of failure rate by $\hat{\lambda} = n / \Delta t$.

Also, the fault birth rate v can be reduced through some information controls such as standardized resource registering, synchronic resource updating, consistent resource descriptions etc, so that to improve the reliability of the resource management system.

■

7.3. Grid Reliability of the Network

If the resource management system has informed the programs of the sites of their required resources in the grid after matchmaking, the running programs are able to access to those resources through the grid network as depicted by the previous Fig. 7.2. Then, the grid program/system/service reliability is mainly determined by the reliability of network, which will be studied in the following subsections.

7.3.1. Reliability model for the grid network

To analyze the grid reliability, two assumptions about the model are given below:

1) The failures of virtual nodes and virtual links can be modeled by Poisson processes.
2) The failures of different elements (nodes and links) are independent from each other.

The first assumption can be justified as in the operational phase without debugging process, the failure rates can remain constant, see e.g. Yang & Xie (2000). The second assumption can be explained as that since the grid is a wide-area distributed system, the nodes and links should be allocated far away from each other so that the possibility of correlation among them can be viewed as very slight or even negligible.

Different programs can exchange information of different sizes with the same resources. Denote by D_{mh} the size of information exchanged between program P_m (m=1,2,...,M) and resource R_h (h=1,2,...,H). The communication time $T_c(i,j)$ between node i and node j, can be derived from

$$T_c(i,j)=\frac{D(i,j)}{S(i,j)} \tag{7.1}$$

where $D(i,j)$ is the total size of information exchanged through the $L(i,j)$, and $S(i,j)$ is the expected bit rate of the link.

Denote the failure rate of the node n by λ_n and of the link $L(i,j)$ by $\lambda_{i,j}$. If any failure occurs either on the link or on the connected two nodes during the communication, the communication process is viewed as a failed process. The reliability of communication between node i and node j through the link $L(i,j)$ can be expressed as

$$R_c(i,j)=\exp\{-(\lambda_i+\lambda_j+\lambda_{i,j})T_c(i,j)\} \tag{7.2}$$

Similarly, during the execution of a program, any failure occurring on the virtual node that executes the program will also make the program failed. The reliability of the node n to run the program P_m, is then given by

$$R_p(m,n)=\exp\{-\lambda_n t(m)\} \tag{7.3}$$

This network reliability model is much more reasonable for the grid than that of conventional distributed systems shown in Chapter 6. Those conventional models somehow inherit the assumptions of Kumar (1986) model. The most stringent assumption that is not suitable for the grid is that the operational probabilities of nodes or links are assumed constant, i.e. $R_c(i, j)$ and $R_p(m,n)$ in the above two equations are constant no matter how long or how different the $T_c(i, j)$ and $t(m)$ are.

Some concepts of grid reliability are defined as follows.

Definition 7.1. *Grid program reliability* (GPR) is defined as the probability of successful execution of a given program running on multiple virtual nodes and exchanging information through virtual links with the remote resources, under the environment of grid computing system.

Then, the grid system reliability (GSR) can be defined as the probability for all of the programs involved in the considered grid system to be executed successfully.

Furthermore, a grid service is to complete certain programs by using some resources distributed in the grid. The grid service reliability is similar to the grid system reliability by considering the programs of the given service, i.e. without taking other programs that are not used by the service into account. Thereby, the grid service reliability is defined as the probability that all the programs of a given service are achieved successfully.

7.3.2. Reliability of minimal resource spanning tree

Recall that the set of virtual nodes and virtual links involved in running the given programs and exchanging information with the resources form a resource spanning tree. The smallest dominating resource spanning tree (*RST*) is called *MRST* (Minimal Resource Spanning Tree). The reliability of an *MRST* is the probability for the *MRST* to be operational to execute the given program. The reliability of an *MRST* denoted by R_{MRST} has three parts:

1) Reliability of all the links contained in the *MRST* during the communication.
2) Reliability of all the nodes contained in the *MRST* during the communication.
3) Reliability of the root node that executes the program during the processing time of the program.

The reliability of the link *L*(*i*, *j*) for exchanging the information can be expressed by

$$R_L(i, j) = \exp\{-\lambda_{i,j} T_c(i, j)\} \tag{7.4}$$

The total communication time of the node G_j can be calculated by

$$T(j) = \sum_{i \in D_j} T_c(i, j) \tag{7.5}$$

where D_j represents the set of nodes that communicate with the node G_j in the *MRST*. The reliability function of the node G_j for communication is

$$R_c(j) = \exp\{-\lambda_j T(j)\} \tag{7.6}$$

Finally, the reliability for a program P_m to be executed successfully during the processing time $t(m)$ on the node *n* is $R_p(m,n)$.

The reliability of the *MRST* can be derived from the above equations as

$$R_{MRST} = R_p(m,n) \prod_{L(i,j) \in MRST} R_L(i, j) \prod_{G_j \in MRST} R_c(j)$$

$$=\exp\{-\lambda_n t(m)\} \prod_{L(i,j)\in MRST} \exp\{-\lambda_{i,j} T_c(i,j)\} \prod_{G_j\in MRST} \exp\{-\lambda_j T(j)\}$$

$$=\exp\{-\lambda_n [t(m)+T(n)]\} \prod_{L(i,j)\in MRST} \exp\{-\lambda_{i,j} T_c(i,j)\} \prod_{\substack{G_j\in MRST \\ j\neq n}} \exp\{-\lambda_j T(j)\} \quad (7.7)$$

In order to simplify the expression, we generalize the term of "communication time" for the root node that contains not only the time of exchanging information with other elements but also the time of executing the given program, i.e. $t(m)+T(n)$.

The term of "*element*" is defined here to represent both the nodes and links of the *MRST*. Assume there are totally *K* elements in an *MRST*, so that $element_i$ (i=1,2,...,K) denotes the *i*:th element in the *MRST*. Accordingly, the communication time of the *i*:th element is denoted by $T_w(element_i)$ and $\lambda(element_i)$ represents its failure rate. The reliability of the *MRST* of the above equation can be simply expressed as

$$R_{MRST} = \prod_{i=1}^{K} \exp\{T_w(element_i)\cdot \lambda(element_i)\} \quad (7.8)$$

With this equation, the reliability of an *MRST* can be computed if the communication time and failure rate of all the elements are given. Hence, finding all the *MRST*s and determining the communication time of their elements are the first step in deriving the grid program reliability and grid system reliability.

The same program executed by different root nodes may cause different communication time on the same elements. Hence, the *MRSTs* should be treated distinctly for the same program executed by different nodes. An example is given below.

Example 7.2. As shown in Fig. 7.5, program P1 can run successfully when either computing node G1 or G4 is successfully working during the processing time, and it is able to successfully exchange information with the required resources (say R1, R2 and R3).

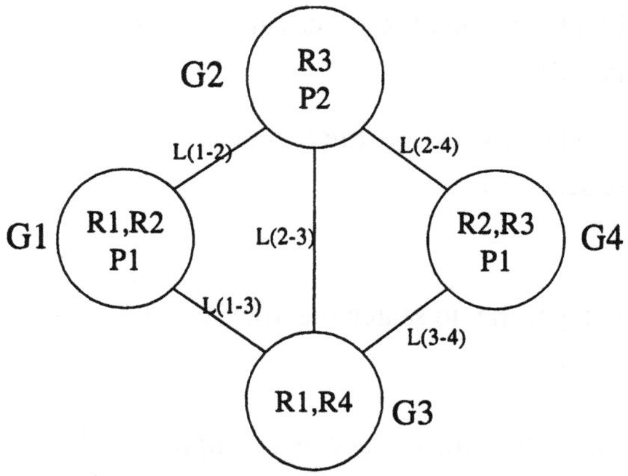

Fig. 7.5. A four-node computing system.

The *MRSTs* considering the communication time of the elements should be separated into two parts:

(a) P1 being executed by G1 contains three *MRSTs*: 1) {G1, G2, L(1,2)}; 2) {G1,G2,G3, L(1,3); 3} {G1,G3,G4,L(1,3),L(3,4)}.

(b) P1 being executed by G4 contains another three *MRSTs*: 4) {G3, G4, L(3,4)}; 5) {G2, G3, G4, L(2,4), L(2,3)}; 6) {G1,G2,G4,L(1,2),L(2,4)}

■

An algorithm is presented in Dai *et al.* (2002) to search the *MRSTs* for a given program executed by one given virtual node. Repeatedly using this algorithm, all the *MRSTs* for different virtual nodes to execute this program can be found, respectively. This algorithm can be briefly described as follows:

Step 1. Start from the given node to search the required resources along the possible links, and record elements that compose the searching route and their communication times.

Step 2. Until all the required resources are reached, an *MRST* is found, and record this *MRST*.

Step 3. Then other routes are tried to search other *MRSTs* until all the *MRSTs* are searched.

An example of the algorithm to search the *MRSTs* is illustrated below.

Example 7.3. Continued with the above Example 7.2. Referring to Fig. 7.5 again, the program P_1 is assumed to exchange information with resources R1,R2,R3 (corresponding exchanged information size are: 500,400,300 Kbit). The bit rates of links L(1,2), L(1,3), L(2,3), L(2,4), L(3,4) are assumed 30, 20, 40, 50, 45 (Kbit/s). Then, search the *MRSTs* for P_1 executed by the node G1 and compute the communication time of each elements in those *MRSTs*, as shown by Fig. 7.6.

Three *MRSTs* are found by the algorithm marked by ☺ in the Fig. 7.6 where all the values in vector RV are 0. The corresponding elements contained in those *MRSTs* are recorded in vector EV with the value 1 and the corresponding communication time is saved in vector WV.

Similarly, other three *MRSTs* for P_1 executed by the other node G4 can also be obtained as listed in the above Example 7.2.

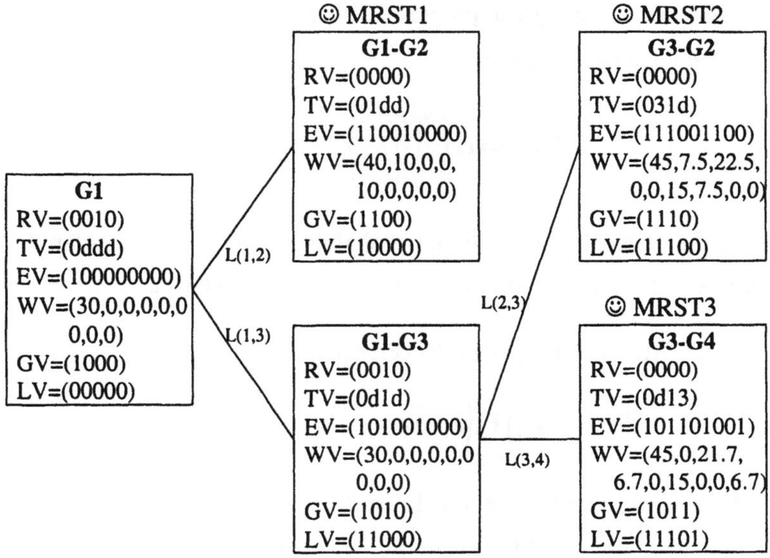

Fig. 7.6. Searching the MRSTs of P1 executed by G1.

■

7.3.3. Grid program and system reliability

Grid program reliability

Note that failures of all the *MRST*s will lead to the failure of the given program, and any one of the *MRST*s can successfully complete the program only if all of its elements are reliable. The grid program reliability of a given program can be described as the probability of having at least one of the *MRST*s working successfully,

$$R(P_m) = \Pr(\text{at least one } MRST \text{ of a given program } P_m \text{ is reliable})$$

Let $N_t(P_m)$ be the total number of *MRST*s for the given program of P_m and E_j be the event in which the $MRST_j$, j=1,2,..., $N_t(P_m)$, is able to successfully execute the given program. The grid program reliability of a given program P_m can be written as

$$R(P_m) = \Pr\left\{ \bigcup_{j=1}^{N_t(P_m)} E_j \right\} \tag{7.9}$$

By using the concept of conditional probability, the events considered in this equation can be decomposed into mutually exclusive events as

$$R(P_m) = \Pr(E_1) + \Pr(E_2)\Pr(\overline{E}_1 | E_2) \ldots + \\ \Pr\{E_{N_t(P_m)}\}\Pr\left\{\overline{E}_1, \overline{E}_2, \cdots, \overline{E}_{N_t(P_m)-1} \middle| E_{N_t(P_m)}\right\} \tag{7.10}$$

where $\Pr(\overline{E}_1|E_2)$ denotes the conditional probability that $MRST_1$ is in the failure state given that $MRST_2$ is in the successful state.

Hence, the grid program reliability can be evaluated in terms of the probability of two distinct events. The first event indicates that the $MRST_i$ is in the operational state while the second indicates that all of its previous trees $MRST_j$ (j=1,2,...,i-1) are in the failure state given that $MRST_i$ is in the operational state. The probability of the first event, $\Pr(E_i)$ is straightforward, and it can be calculated through Eq. (7.8). The probability of the second event, $\Pr(\overline{E}_1, \overline{E}_2, \cdots, \overline{E}_{i-1}|E_i)$, can be computed using the algorithms presented by Dai *et al.* (2002).

The brief introduction of the algorithm is given here. It contains two steps.

Step 1 identifies all the conditional elements that can lead to the failure of any $MRST_j$ (j=1,2,...,$i-1$) while keeping $MRST_i$ to be operational. Such a conditional element, say $element_k$ (contained in any $MRST_j$, j=1,2,...,$i-1$), has starting time and end time. If any failure occurs on the $element_k$ between its starting time and end time, it can lead the $MRST_j$ to fail.

Step 2 uses a binary search tree (Johnsonbaugh, 2001: pp. 349-354) to seek the possible combinations of these identified elements that can make all the $MRST_j$ (j=1,2,...,$i-1$) fail and computes the probabilities of those combinations.

The summation of the probabilities is the result of $\Pr(\overline{E}_1, \overline{E}_2, \cdots, \overline{E}_{i-1} | E_i)$. For detailed procedures of the two steps can be found in Dai *et al.* (2002). An example of this algorithm can also be found.

Grid system reliability

The grid system reliability equation can be written as the probability of the intersection of the set of *MRST*s of each program, which is

$$R_S = \Pr\left\{ \bigcap_{m=1}^{M} MRST(P_m) \right\} \tag{7.11}$$

where $MRST(P_m)$ denotes the set of all the *MRST*s associated with the program P_m.

The intersection of the trees of each $MRST(P_m)$ can be evaluated first by intersecting $MRST(P_1)$. The intersected tree of two *MRST*s is generated by putting all the elements of the two *MRST*s together, where the communication time of overlapped elements should be added together. An example of intersected *MRST* is illustrated below.

Example 7.4. Suppose one *MRST* related to program P_1 is {G1,G2,G3,L(1,3),L(2,3)} with the communication time {45, 7.5, 22.5, 15, 7.5} and one *MRST* related to program P_2 is {G1,G2,G3,L(1,2),L(1,3)} with the communication time {50, 70, 30, 20, 30}. Then, the intersected *MRST* of the above two *MRSTs* should be {G1,G2,G3,L(1,2),L(1,3),L(2,3)} with the communication time {95, 77.5, 52.5, 20, 45, 7.5}.

■

In fact, if any one of the intersected *MRSTs* of $MRST(P_m)$ (m=1,2,...,M) is reliable, all the programs required in the grid system can be successfully completed; If all the intersected *MRSTs* fail, the grid system cannot be successfully completed.

After generating all the intersected *MRSTs*, the grid system reliability can be written as

$$R_S = \Pr\left(\bigcup_{j=1}^{N_t} E_j\right) \tag{7.12}$$

where N_t is the total number of intersected *MRSTs*. This equation is similar to the previous Eq. (7.9), so the above algorithms for deriving the grid program reliability can be similarly used in deriving the grid system reliability here.

Grid service reliability

The grid service reliability can be viewed as a special type of the grid system reliability if we consider the grid service in a way that the whole grid system is only providing this required service and other services are not considered now. With this classification, the concept of grid system reliability is generalized to include the reliability of different number of services.

All the above algorithms computing the grid program/system reliability are illustrated by a numerical example as below, and then the reliability of resource management system is also integrated into the grid reliability analysis.

Example 7.5. Suppose that a simple grid system is to provide a web service of "Stock Analysis" for different countries. Three different resources (R1,R2,R4) store the real-time stock price of different countries, and another resource (R3) is the database of a website that outputs and shows the results out of the "Stock Analysis". The service procedure can be described as that two programs (P1 and P2) collect data from the three resources (R1,R2,R4) to analyze the stock market information for different countries, and then output the results into the database (R3) which can be loaded by a website service.

Revisit Fig. 7.5 that contains four virtual nodes and five virtual links and runs the two programs and prepare the four resources. Tables 7.1-7.2 show the necessary input information.

Table 7.1. Failure rate and speed of elements (links and nodes).

Elements	L(1,2)	L(1,3)	L(2,3)	L(2,4)	L(3,4)	G1	G2	G3	G4
Failure rate	0.001	0.002	0.003	0.004	0.005	0.001	0.0001	0.003	0.004
Speed (Kbps)	30	20	40	50	45				

Table 7.2. processing time and information exchanged with the resources.

Program	Run Time (Sec)	Resources	Exchanged information (Kbit)
P1	30	R1, R2, R3	500,400,300
P2	50	R3, R4	200,600

With the approaches presented above, Table 7.3 shows all *MRST*s of the program P1 with the communication time of each element evaluated by the above Example 7.3 and its reliability $\Pr(E_i)$ calculated by Eq. (7.8). Table 7.3

also shows the conditional probability of p= $\Pr(\overline{E}_1,\overline{E}_2,\cdots,\overline{E}_{i-1}|E_i)$ evaluated similarly as the above Example 7.4.

Table 7.3. Evaluation for the grid program reliability of P1.

MRST$_i$	Elements	Communication time (T_w)	$\Pr(E_i)$	p
$MRST_1$	G1,G2,L(1,2)	40,10,10	0.950279	----------
$MRST_2$	G1,G2,G3,L(1,3),L(2,3)	45,7.5,22.5,15,7.5	0.847258	0.010198
$MRST_3$	G1,G3,G4,L(1,3),L(3,4)	45,21.7,6.7,15,6.7	0.818403	0.001000
$MRST_4$	G3,G4,L(3,4)	11.1,41.1,11.1	0.776313	0.039890
$MRST_5$	G2,G3,G4,L(2,4),L(2,3)	22.5,12.5,40,10,12.5	0.755973	0.002314
$MRST_6$	G1,G2,G4,L(1,2),L(2,4)	16.7,26.7,40,16.7,10	0.789725	0.000810

Substituting the values of $\Pr(E_i)$ and $\Pr(\overline{E}_1,\overline{E}_2,\cdots,\overline{E}_{i-1}|E_i)$ of Table 7.3 into Eq. (7.10), the grid program reliability of P1 is

$$R(P1)= 0.99309$$

Similarly, the grid program reliability of P2 can be obtained as

$$R(P2)=0.773368+0.769665\times 0.122694+0.90801\times 0.0907=0.950158$$

where three *MRST*s are found for P2 to be executed by G2.

The grid system reliability can then be derived. The total number of intersected trees is $6\times 3=18$. Similar to grid program reliability, the grid system reliability is obtained as

$$R_S=0.926380$$

Suppose that the total time for resource management system to deal with the program P1's requests is t=15 seconds and the failure rate at that time slot $\lambda = 0.0005$ (sec^{-1}). The reliability for the request of the program P1 is then computed as

$$R_{RMS}(P1) = \exp(-\lambda t) = 0.992528$$

The grid program reliability of P1 considering the reliability of resource management system can be calculated by multiplying the above R_{RMS} together with $R(P1)$ as

$$GPR(P1) = R_{RMS}(P1) \cdot R(P) = 0.992528 \times 0.99309 = 0.98567$$

For P2, if the total time for resource management system to deal with its resource requests is 10 seconds, a similar way can be used to obtain

$$R_{RMS}(P2) = \exp(-0.0005 \times 10) = 0.99501$$

Multiplying it with $R(P2)$, we get

$$GPR(P2) = R_{RMS}(P2) \cdot R(P2) = 0.99501 \times 0.950158 = 0.94542$$

For the grid system reliability that includes both P1 and P2, the reliability can be computed as

$$GSR = R_{RMS}(P1) \cdot R_{RMS}(P2) \cdot R_s = 0.992528 \times 0.99501 \times 0.926380 = 0.91487$$

■

7.4. Grid Reliability of the Software and Resources

In the above section, the grid reliability is analyzed by considering only the network hardware failures, i.e. failures of processing nodes and communicating links. However, software program failures and resource failures should also be integrated into the grid reliability analysis.

7.4.1. Reliability of software programs and resources

Besides hardware causes, failures of a software program may also be caused by the faults in the program itself. In the operational phase, the software program failures can be assumed to follow the exponential distributions here. The software failure occurrence rate of program P_i running on processing node G_j is denoted by $\lambda_s(i,j)$, because a same program running on different processing nodes may have different failure rates. Also, the processing time of P_i on G_j is denoted by $t(i,j)$. Thus, the reliability of the software program P_i running on G_j can be simply computed by

$$R_{\text{prog}}(i,j) = \exp\{-\lambda_s(i,j) \cdot t(i,j)\} \tag{7.13}$$

For the resource reliability, the previous section assumes that if the program uses the resource, the resource itself is perfect and the failures only occur when transferring the information through the communication network. However, the resource possibly risks failures when it is needed.

Suppose the time for resource *h* to work is determined by the program P_i by which the resource is requested and the node G_j on which the resource is integrated, denoted by $t(h,i,j)$. Also, considering the operational phase for the integrated resources, we denote the failure rate of the resource *h* on the node G_j by $\lambda_r(h,j)$, which follows the exponential distribution. Thus, the reliability of resource *h* requested by P_i and integrated on G_j can be simply expressed by

$$R_{\text{res}}(h,i,j) = \exp\{-\lambda_r(h,j) \cdot t(h,i,j)\} \tag{7.14}$$

7.4.2. Grid reliability integrating software and resource failures

In order to integrate the software program and resource failures into grid reliability analysis together with the hardware network reliability, we revise the model presented in Section 7.3. For each virtual node, consider its programs and resources as its sub nodes, as shown by Fig. 7.7. Here G_j is a virtual node on which $P_{x1},\ldots,P_{xm}$ are attached as the sub nodes representing programs and $R_{y1}\ldots R_{yk}$ corresponding to resources.

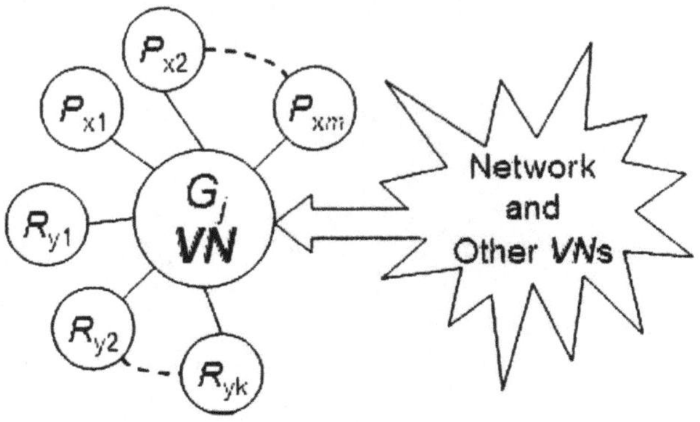

Fig. 7.7. Virtual node and its sub nodes of programs and resources.

Such abstraction of the Fig. 7.7 has the following advantages:

1) The reliability of different software programs and resources can be integrated into the grid reliability analysis given the failure rates of all the sub nodes and their communication time.

2) It incorporates the hardware reliability in the grid reliability analysis and the common cause failures among those programs and resources are considered. For example, if G_j fails, all its sub nodes (corresponding to the programs or resources executed by or integrated on the same virtual node) are no longer working.

3) All the approaches presented in the Section 7.3 can be directly implemented to compute grid program/system/service reliability if each sub node is viewed as an element itself, and the link between the virtual node and its sub node is assumed to be perfect.

Example 7.6. Revisit Fig. 7.5. Replace the nodes with those in the Fig. 7.7 that considers the software program and resource failures. Fig. 7.8 depicts the new network graph for the grid computing system containing the sub nodes of programs and resources. The approaches presented by Section 7.3 can be directly and similarly implemented in deriving the grid reliability of Fig. 7.8.

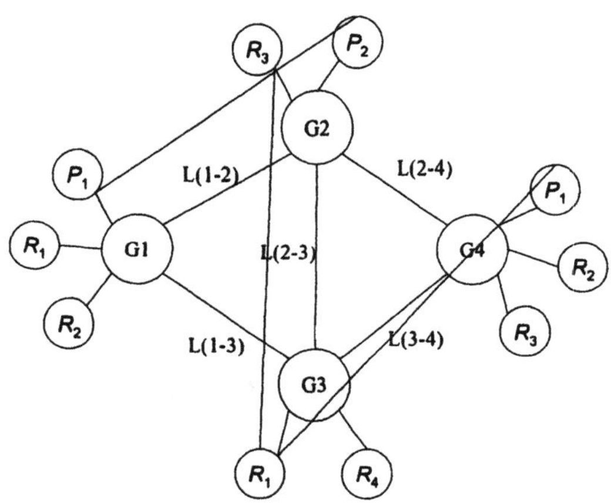

Fig. 7.8. Grid network containing the sub nodes of programs and resources.

■

7.5. *Notes and References*

Foster & Kesselman (1998) summarized the basic concepts of the grid and presented a grid development tool which addresses issues of security, information discovery, resource management, data management, communication, and portability. It is implemented in many Grid projects. Recently, Foster *et al.* (2002) further developed the grid technologies toward an

Open Grid Services Architecture in which a Grid provides an extensible set of services that virtual organizations can aggregate in various ways.

For the resource management systems, Krauter *et al.* (2002) classified the existing techniques into different types according to their control property and investigated their applications. In order to address complex resource management issues such as cost, Buyya *et al.* (2002) further proposed a computational economy framework for resource allocation and for regulating supply and demand in the gird computing environments. This framework provides mechanisms for optimizing resource provider and consumer objective functions through trading and brokering services. Cao *et al.* (2002) also presented an agent-based resource management system that was implemented for the grid computing. It utilized the performance prediction techniques of the PACE toolkit to provide quantitative data regarding the performance of complex applications running on a local grid resource.

For the network issues of the grid, Postel & Touch (1998) reviewed the evolution of network techniques in different stages and summarized those that could be implemented into the grid network. Keahey *et al.* (2002) introduced the concept of "network services" in their "National Fusion Collaboratory" project, which build on the top of the computational grids, and provide Fusion codes, together with their maintenance and hardware resources as a service to the community. Weissman & Lee (2002) also presented the design of system architecture, called Virtual Service Grid, for delivering high-performance network services.

CHAPTER 8

MULTI-STATE SYSTEM RELIABILITY

Most of reliability models for computing systems assume only two possible states of the system: operational state and failed state. In reality, many systems exhibit noticeable gradations of performance besides the above two. For example, if some computing elements in a computing system fail, the system may still continue working but its performance may be degraded. Such degradation state is another state between the perfect working state and the completely failed state. To study this type of systems, the multi-state system (MSS) reliability is investigated in this chapter.

The chapter is divided into three parts. First, the basic concepts of the MSS are introduced. Some basic Markov models for MSS reliability analysis are then presented. Finally, the MSS failure correlation model is studied using a Markov renewal process model.

8.1. Basic Concepts of Multi-State System (MSS)

All engineering systems are designed to perform their intended tasks in a given environment. Some systems can perform their tasks with various distinguished levels of efficiency which can be referred to as *performance levels.* A system that can have a finite number of performance levels is referred to as a multi-state system, e.g. Brunelle & Kapur (1999), Pourret *et al.* (1999), Lisnianski & Levitin (2003) and Wu & Chan (2003).

A binary system is the simplest case of the MSS having only two distinguished states. There are many different situations in which a system should be considered to be a MSS:

1) Any system consisting of different units that have a cumulative effect on the entire system performance can be considered as a MSS.
2) The performance level of elements composing a system can also vary as a result of their deterioration (fatigue, partial failures) or because of variable ambient conditions.

8.1.1. Generic MSS model

A system element j is assumed to have k_j different states of the performance level, represented by the set

$$g_j = \{g_{j1}, g_{j2}, ..., g_{jk_j}\}$$

where g_{ji} denotes the performance level of element j in the state i, $i \in \{1,2,...,k_j\}$. The performance level $G_j(t)$ of element j at any instant $t \geq 0$ is a random variable that takes its values from g_j: $G_j(t) \in g_j$. Therefore, for the time interval $[0,T]$, where T is the MSS operation period, the performance level of element j is defined as a stochastic process (Lisnianski & Levitin, 2003).

In some cases, the element performance cannot be measured only by a single value, but by more complex mathematical objects, usually vectors. In these cases, the element performance is defined as a vector stochastic process $G_j(t)$.

The probabilities associated with the different states (performance levels) of the system element j at any instant t can be represented by the set

$$p_j(t) = \{p_{j1}(t), p_{j2}(t), \ldots, p_{jk_j}(t)\} \tag{8.1}$$

where

$$p_{ji}(t) = \Pr\{G_j(t) = g_{ji}\} \tag{8.2}$$

Note that since the states of an element compose the complete group of mutually exclusive events, we have

$$\sum_{i=1}^{k_j} p_{ji}(t) = 1,\ 0 \le t \le T$$

Eq. (8.2) defines the mass function $G_j(t)$ for discrete performance levels at any instant t. The collection of pairs g_{ji}, $p_{ji}(t)$, $i = 1,2,\ldots,k_j$, completely determines the probability distribution of performance of the element j at any instant t, see, e.g., Lisnianski & Levitin (2003).

When the MSS consists of n elements, its performance levels are unambiguously determined by the performance levels of these elements. At any time, the system elements have certain performance levels corresponding to their states. The state of the system has K different states and that g_i is the entire system performance level in state i, $i \in \{1,\ldots,K\}$. The MSS performance level at time t is a random variable that takes values from the set $\{g_1,\ldots,g_K\}$.

Let

$$L^n = \{g_{11},\ldots,g_{1k_1}\} \times \{g_{21},\ldots,g_{2k_2}\} \times \cdots \times \{g_{n1},\ldots,g_{nk_n}\}$$

be a space of possible combinations of performance levels for all of the system elements and $M = \{g_1,\ldots,g_K\}$ is a space of possible values of the performance level for the entire system. The transform

$$\phi(G_1(t),\ldots,G_n(t)):\ L^n \to M$$

which maps the space of the elements' performance levels into the space of system performance levels, is called the system structure function. Note that the MSS structure function is an extension of a binary structure function. The only difference is in the definition of the state spaces: the binary structure function is mapped as $\{0,1\}^n \to \{0,1\}$, while in the MSS, one deals with more complex spaces.

A generic model of the multi-state system can be defined as follows. The performance processes are modelled as stochastic process $G_j(t)$, j=1,2,...n, for each system element j. The system structure function that produces the stochastic process corresponding to the output performance of the entire MSS is

$$G(t) = \phi(G_1(t), G_2(t), \ldots, G_n(t))$$

In practice, a simpler MSS model can be used. This can be based on probability distribution of performances for all of the system elements at any instant time t during the operation period $[0,T)$ and system structure function:

$$g_j(t), p_j(t), \; 1 \le j \le n \tag{8.3}$$

and

$$\phi(G_1(t), G_2(t), \ldots, G_n(t)) \tag{8.4}$$

The system state can also be represented in a table, in analytical form, or be described as an algorithm for unambiguously determining the system performance $G(t)$ for any given set $\{G_1(t), G_2(t), \ldots, G_n(t)\}$. An example of MSS modeling is illustrated below.

Example 8.1. Consider a 2-out-of-3 MSS. This system consists of 3 binary elements with the performance levels $G_i(t) \in \{g_{i1}, g_{i2}\} = \{0,1\}$, for i=1,2,3, where

$$g_{i1} = \begin{cases} 0, & \text{if the element i is in the state of complete failure;} \\ 1, & \text{if the element i functions perfectly.} \end{cases}$$

The system output performance level $G(t)$ at any instant t is:

$$G(t) = \begin{cases} 0, & \text{if there is more than one failed element;} \\ 1, & \text{if there is only one failed element;} \\ 2, & \text{if all the elements function perfectly.} \end{cases}$$

The values of the system structure function $G(t) = \phi(G_1(t), G_2(t), G_3(t))$ for all the possible system states are presented in Table 8.1.

Table 8.1: Structure function for 2-out-of-3 system

$G_1(t)$	$G_2(t)$	$G_3(t)$	$\phi(G_1(t), G_2(t), G_3(t))$
0	0	0	0
0	0	1	0
0	1	0	0
0	1	1	1
1	0	0	0
1	0	1	1
1	1	0	1
1	1	1	2

■

8.1.2. Basic MSS reliability measures

To characterize MSS behavior from a reliability point of view one has to determine the MSS reliability measures. These measures can be considered as extensions of the corresponding reliability measures for a binary-state system. Brunelle & Kapur (1999) and Lisnianski & Levitin (2003) reviewed many MSS reliability measures. Some commonly used ones are introduced as follows.

Since the system is characterized by its output performance $G(t)$, the state acceptability depends on the value of this index. This dependency can be expressed by the acceptability function $F(G(t))$ that takes non-negative values if and only if the MSS functioning is acceptable. This takes place when the efficiency of the system functioning is completely determined by its internal state

(Lisnianski & Levitin, 2003). In such cases, a particular set of MSS states is of interest to the customer. Usually these states are interpreted as system failure states, which when reached, imply that the system should be repaired or discarded. The set of acceptable states can be also defined when the system functionality level is of interest at a particular point in time (such as at the end of the warranty period).

More frequently, the system state acceptability depends on the relation between the MSS performance and the desired level of this performance (demand). In general, the demand $W(t)$ is also a random process. It can take discrete values from the set $w = \{w_1, ..., w_M\}$. The desired relation between the system performance and the demand can also be expressed by the acceptability function $F(G(t), W(t))$. The acceptable system states correspond to $F(G(t), W(t)) \geq 0$ and the unacceptable states correspond to $F(G(t), W(t)) < 0$. The last inequality defines the MSS failure criterion.

The performance of MSS should exceed the demand. In such cases the acceptability function takes the form

$$F(G(t), W(t)) = G(t) - W(t)$$

Since $G(t)$ and $W(t)$ are random processes, the subset of acceptable states can vary in time. The system behavior during the operation period can be characterized by the possibility of entering the subset of unacceptable states more than once. The case when MSS can enter this subset only once corresponds to non-repairable systems. For repairable systems or for the systems with variable demands, the transitions between subsets of acceptable and unacceptable states may occur an arbitrary number of times.

Some other reliability measures are based on the above acceptability function $F(G(t), W(t))$. The following random variables can be of interest:

(a) Time to failure, T_f, is the time from the beginning of the system life up to the instant when the system enters the subset of unacceptable states the first time.

(b) Time between failures, T_b, is the time between two consecutive transitions from the subset of acceptable states to the subset of unacceptable states.

(c) Number of failures, N_T, is the number of times the system enters the subset of unacceptable states during the time interval $[0,T]$.

The probability of a failure-free operation or reliability function is

$$R(t) = \Pr\{T_f \geq t \mid F(G(0), W(0)) \geq 0\}$$

The Mean Time To Failure (MTTF) is the expected time up to the instant when the system enters the subset of unacceptable states for the first time, as $E(T_f)$.

The MSS instantaneous (point) availability $A(t)$ is the probability that the MSS at instant t is in one of the acceptable states:

$$A(t) = \Pr\{F(G(t), W(t)) \geq 0\}$$

The MSS availability in the time interval $[0,T]$ is defined as:

$$A_T = \frac{1}{T}\int_0^T A(t)dt \tag{8.5}$$

which represents the portion of time when the MSS output performance level is in an acceptable area.

Wu & Chan (2003) further presented an MSS measure called expected utility function to evaluate the overall performance of the MSS at a time instant t, expressed by

$$U(t) = \sum_{j=0}^{K} a_j \Pr\{G(t) = j\}$$

where a_j is the utility of the MSS to stay at state j.

8.2. Basic Models for MSS Reliability

According to the generic MSS model, any system element j can have k_j different states corresponding to the performance levels, represented by the set $g_j = \{g_{j1},...,g_{jk_j}\}$. The current state of the element j and, therefore, the current value of the element performance level $G_j(t)$ at any instant t, are random variables. $G_j(t)$ takes values from g_j : $G_j(t) \in g_j$. Therefore, for the time interval $[0,T]$, where T is the MSS operation period, the performance level of element j is defined as a stochastic process.

In this section when we deal with a single multi-state element, the index j will be omitted for the designation of a set of the element's performance levels. This set is denoted as $g = \{g_1,...,g_k\}$. We also assume that this set is ordered so that $g_{i+1} \geq g_i$ for any i.

8.2.1. Non-repairable multi-state elements

The lifetime of a non-repairable element lasts until its first entrance into the subset of unacceptable states. In general, the acceptability of the state of an element depends on the relation between the performance of the element and the desired level of this performance (demand). The demand $W(t)$ is also a random process that takes discrete values from the set $w = \{w_1,...,w_M\}$. The desired relation between the system performance and the demand can be expressed by the acceptability function $F(G(t),W(t))$.

Minor failures

First consider a multi-state element with only minor failures defined as failures that cause element transition from state i to the adjacent state i-1. In other words, minor failure causes minimal degradation of element performance. The CTMC for such an element is presented in Fig. 8.1.

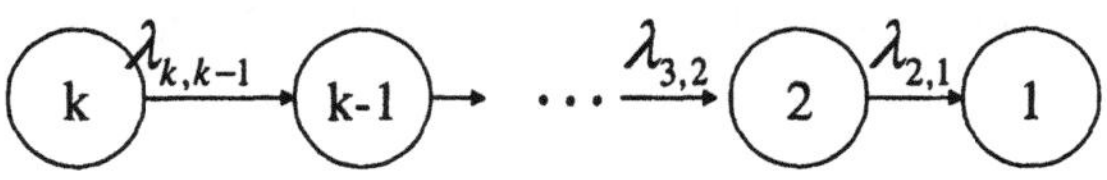

Fig. 8.1. CTMC for non-repairable element with minor failures.

Denote by $P_i(t)$ (i=1,2,...k) the probability for the system to stay at state i at time instant t. Then the Chapman-Kolmogorov equation can be written as

$$P_k'(t) = -\lambda_{k,k-1} P_k(t)$$

$$P_i'(t) = \lambda_{i+1,i} P_{i+1}(t) - \lambda_{i,i-1} P_i(t), \quad i=2,3,\ldots,k-1 \tag{8.6}$$

$$P_1'(t) = \lambda_{2,1} \cdot P_2(t)$$

We assume that the process starts from the best state k with a maximal element performance level of g_k. Hence, the initial conditions are

$$P_k(0) = 1$$

and

$$P_{k-1}(0) = P_{k-2}(0) = \ldots = P_1(0) = 0$$

One can obtain the numerical solution of the above differential equations under the initial conditions even for large k. They can also be solved analytically by using Laplace-Stieltjes transform in some cases. With this transform and by taking into account the initial conditions, one can represent the above differential equations in the form of linear algebraic equations and solved to obtain

$$P_k(s) = \frac{1}{s + \lambda_{k,k-1}}$$

$$P_i(s) = \frac{1}{s + \lambda_{k,k-1}} \cdot \frac{\lambda_{k,k-1}}{s + \lambda_{k-1,k-2}} \cdots \frac{\lambda_{i+1,i}}{s + \lambda_{i,i-1}}, \quad i=2,3,\ldots,k-1 \tag{8.7}$$

$$P_1(s) = \frac{1}{s + \lambda_{k,k-1}} \cdot \frac{\lambda_{k,k-1}}{s + \lambda_{k-1,k-2}} \cdots \frac{\lambda_{3,2}}{s + \lambda_{2,1}} \cdot \frac{\lambda_{3,2}}{s}$$

Now in order to find the function $P_k(t)$, the inverse Laplace-Stieltjes transform can be applied.

The probability of the state with the lowest performance $P_1(t)$ determines the unreliability function of the multi-state element for the constant demand level $g_2 \geq w > g_1$. The reliability function defined as the probability that the element is not in its worst state (total failure) is

$$R_1(t) = 1 - P_1(t) \tag{8.8}$$

In general, if the constant demand is $g_{i+1} \geq w > g_i$ (i=1,...,k-1), the unreliability function for the multi-state element is a sum of the probabilities of the unacceptable states 1,...,i. The reliability function is then

$$R_i(t) = 1 - \sum_{j=1}^{i} P_j(t) \tag{8.9}$$

The mean time up to multi-state element failure for this constant demand level can be interpreted as the time of the process entering state i. It can be calculated as the sum of the time periods during which the process is remaining in each state $j > i$. Since the process begins from the best state k with the maximal element performance level, we have

$$MTTF_i = \sum_{j=i+1}^{k} \frac{1}{\lambda_{j,j-1}} \tag{8.10}$$

Example 8.2. Consider a non-repairable multi-state system that has only minor failures. The system has 4 possible states whose performance levels are set as 100, 80, 50 and 0, respectively. Its Markov model can be built as Fig. 8.1 with k=4. Assume that the failure rates are given by

$$\lambda_{4,3} = 0.02,\ \lambda_{3,2} = 0.01,\ \lambda_{2,1} = 0.007$$

and the initial state is the best state, state 4.

Substituting the above numerical values into the Laplace-Stieltjes transforms and inverting them, the state probabilities can be obtained. The state probabilities as a function of time are shown in Fig. 8.2.

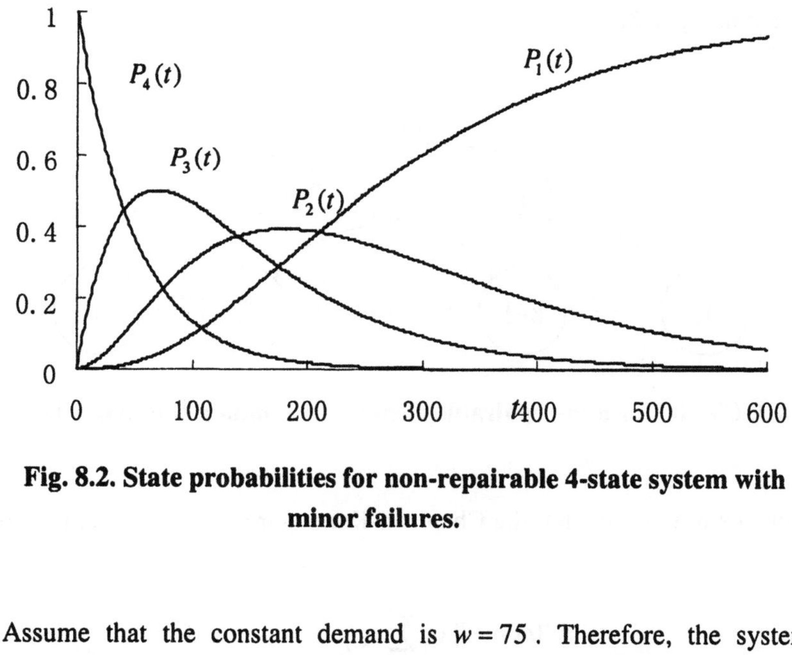

Fig. 8.2. State probabilities for non-repairable 4-state system with minor failures.

Assume that the constant demand is $w = 75$. Therefore, the system is reliable only if the system is at least at state 3 with performance level 80. Then, the reliability function can be obtained as

$$R_2(t) = 1 - P_1(t) - P_2(t)$$

Then, the mean time to failure is obtained by

$$MTTF_2 = \frac{1}{\lambda_{4,3}} + \frac{1}{\lambda_{3,2}} = 150$$

■

Both minor and major failures

Now consider a non-repairable multi-state element that can have both minor and major failures (major failure is a failure that causes the element transition from state i to state j: $j < i-1$). The state-space diagram for such an element representing transitions corresponding to both minor and major failures is presented in Fig. 8.3.

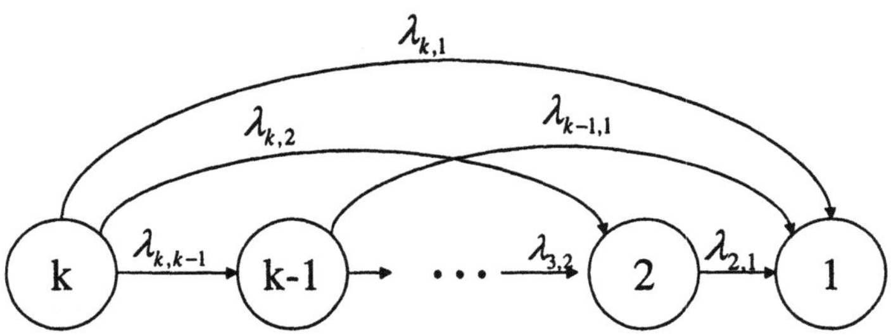

Fig. 8.3. CTMC for non-repairable element with minor and major failures.

For this Markov model, the Chapman-Kolmogorov equation can be written as

$$P_k'(t) = -P_k(t)\sum_{j=1}^{k-1}\lambda_{k,j}$$

$$P_i'(t) = \sum_{j=i+1}^{k}\lambda_{j,i}P_j(t) - P_i(t)\sum_{j=1}^{i-1}\lambda_{i,j}\ , \quad i=2,3,\dots,k\text{-}1 \tag{8.11}$$

$$P_1'(t) = \sum_{j=2}^{k}\lambda_{j,1}P_j(t)$$

After solving the above equations and obtaining the state probabilities $P_i(t)$, the reliability can be easily derived as Eqs. (8.8-8.9).

8.2.2. Repairable multi-state elements

Availability modeling

A more general model of a multi-state element is the model with repair. The repairs can also be both minor and major. The minor repair returns an element from state j to state $j+1$ while the major repair returns it from state j to state i, where $i > j+1$, see, e.g., Lisnianski & Levitin (2003).

A special case is when an element has only minor failures and minor repairs. It is actually a birth and death process. The CTMC of this process is presented in Fig. 8.4. The CTMC for the general case of the repairable multi-state element with minor and major failures and repairs is presented in Fig. 8.5.

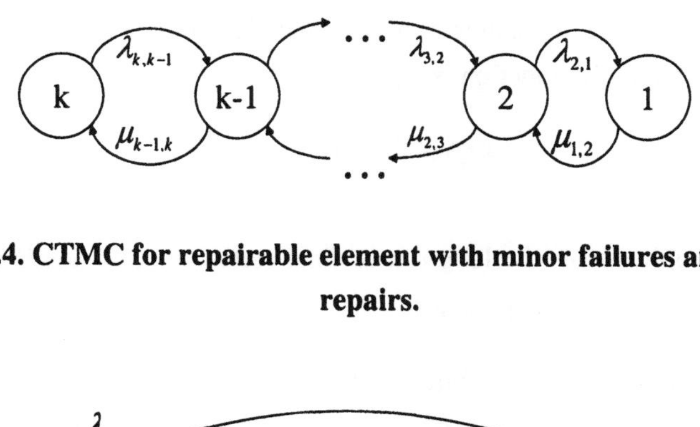

Fig. 8.4. CTMC for repairable element with minor failures and minor repairs.

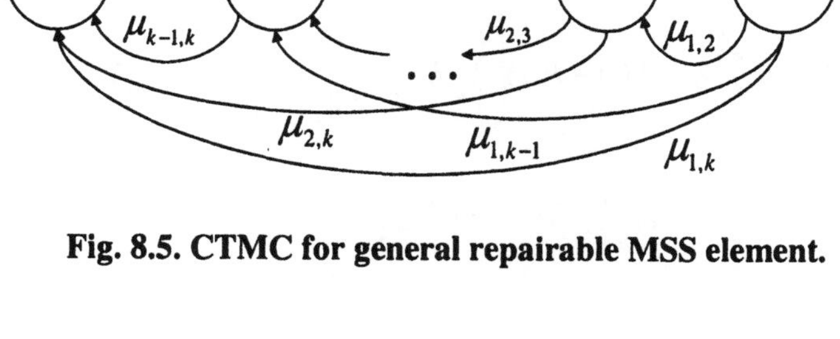

Fig. 8.5. CTMC for general repairable MSS element.

The following are Chapman-Kolmogorov equations for the general case:

$$P_k'(t) = \sum_{j=1}^{k-1} \mu_{j,k} P_j(t) - P_k(t) \sum_{j=1}^{k-1} \lambda_{k,j}$$

$$P_i'(t) = \sum_{j=i+1}^{k} \lambda_{j,i} P_j(t) + \sum_{j=1}^{i-1} \mu_{j,i} P_j(t) - P_i(t) \left(\sum_{j=1}^{i-1} \lambda_{i,j} + \sum_{j=i+1}^{k} \mu_{i,j} \right), \; i=2,3,\ldots,k\text{-}1$$

$$P_1'(t) = \sum_{j=2}^{k} \lambda_{j,1} P_j(t) - P_1(t) \sum_{j=2}^{k} \mu_{1,j} \qquad (8.12)$$

Solving the above equations, one obtains the state probabilities $P_i(t)$ (i=1,2,…,k).

When $F(g_i, w) = g_i - w$ for the constant demand level $g_{i+1} \geq w > g_i$, the acceptable states where the element performance is above level g_i are $i+1$, …, k. Hence, the availability function is

$$A_i(t) = \sum_{j=i+1}^{k} P_j(t) \qquad (8.13)$$

In many applications, the long-run or final states probabilities $\lim_{t \to \infty} P_i(t)$ are of interest for the repairable element.

For the long run state probabilities, the computations become simpler. The above differential equations is reduced to a set of k algebraic linear equations because for the constant probabilities, all time derivatives $P_i'(t) = 0$ as below

$$\sum_{j=1}^{k-1} \mu_{j,k} P_j(t) - P_k(t) \sum_{j=1}^{k-1} \lambda_{k,j} = 0$$

$$\sum_{j=i+1}^{k} \lambda_{j,i} P_j(t) + \sum_{j=1}^{i-1} \mu_{j,i} P_j(t) - P_i(t) \left(\sum_{j=1}^{i-1} \lambda_{i,j} + \sum_{j=i+1}^{k} \mu_{i,j} \right) = 0, \; i=2,3,\ldots,k\text{-}1 \qquad (8.14)$$

$$\sum_{j=2}^{k} \lambda_{j,1} P_j(t) - P_1(t) \sum_{j=2}^{k} \mu_{1,j} = 0$$

An additional independent equation can be provided by the simple fact that the sum of the state probabilities is equal to 1 at any time. The above equations can then be solved.

Example 8.3. Consider a 4-state repairable system with both minor and major failures and repairs. The performance levels of the four states are

$$g_4 = 100, \; g_3 = 80, \; g_2 = 50, \; g_1 = 0$$

respectively. The unit has the following failure rates:

$\lambda_{4,3} = 0.02$, $\lambda_{3,2} = 0.006$, $\lambda_{2,1} = 0.007$, $\lambda_{3,1} = 0.002$, $\lambda_{4,2} = 0.005$, $\lambda_{4,1} = 0.003$

and the following repair rates:

$$\mu_{3,4} = 1, \; \mu_{2,3} = 0.8, \; \mu_{1,2} = 0.5, \; \mu_{1,4} = 0.32, \; \mu_{1,3} = 0.4, \; \mu_{2,4} = 0.45$$

The Markov model can be constructed as Fig. 8.5 with k=4. Substituting the above numerical values into Eq. (8.12), we can obtain the state probability functions as depicted by Fig. 8.6.

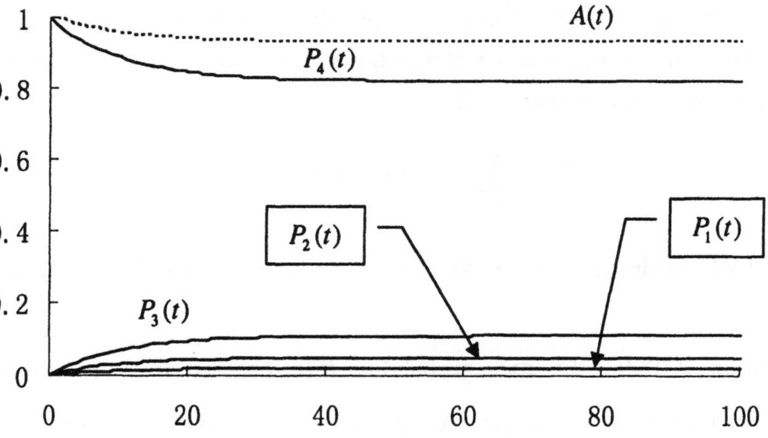

Fig. 8.6. State probabilities of the repairable 4-state system.

Assume that the constant demand $w = 75$. The available states of the system are states 4 and 3, so the system availability function is

$$A(t) = P_4(t) + P_3(t)$$

which is also shown in Fig. 8.6 as the dashed line.

■

Reliability modeling

The determination of the reliability function for the repairable multi-state element is based on finding the probability of the event when the element enters the set of unacceptable states the first time. In order to find the element reliability function $R_i(t)$, for the constant demand w ($g_{i+1} \geq w > g_i$), an additional Markov model should be built. All states $1,2,\ldots,i$ of the element corresponding to the performance levels lower than the demand w, should be combined in one absorbing state. This absorbing state can be considered now as state 0 and all repairs that return the element from this state back to the set of acceptable states should be forbidden.

The transition rate $\lambda_{m,0}$ from any acceptable state m ($m > i$) to the combined absorbing state 0 is equal to the sum of the transition rates from the state m to all the unacceptable states (states $1,2,\ldots,i$):

$$\lambda_{m,0} = \sum_{j=1}^{i} \lambda_{m,j}, \text{ for } m=i+1,\ldots,k \tag{8.15}$$

The CTMC model for the computation of the reliability function is depicted by Fig. 8.7

For this CTMC, the state probability $P_0(t)$ characterizes the unreliability function of the element because after the first entrance into the absorbing state 0 the element never leaves it, i.e., we have

$$R_i(t) = 1 - P_0(t).$$

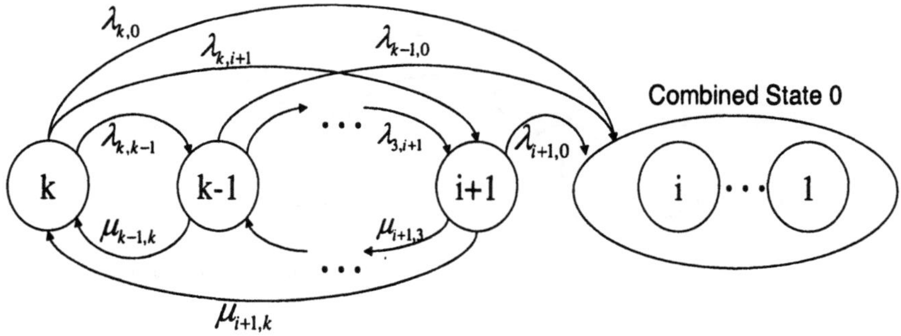

Fig. 8.7. CTMC for determination of reliability function for repairable element.

It is easy to obtain $P_0(t)$ by solving the following Chapman-Kolmogorov equations:

$$P_k'(t) = \sum_{j=i+1}^{k-1} \mu_{j,k} P_j(t) - P_k(t)\left(\sum_{j=i+1}^{k-1} \lambda_{k,j} + \lambda_{k,0}\right)$$

$$P_m'(t) = \sum_{j=m+1}^{k} \lambda_{j,m} P_j(t) + \sum_{j=i+1}^{m-1} \mu_{j,m} P_j(t) - P_m(t)\left(\sum_{j=i+1}^{m-1} \lambda_{m,j} + \sum_{j=m+1}^{k} \mu_{m,j} + \lambda_{m,0}\right),\ i<m<k$$

$$P_0'(t) = \sum_{j=i+1}^{k} \lambda_{j,0} P_j(t) \tag{8.16}$$

The reliability function $R_i(t)$ can then be obtained.

Example 8.4. Continue with Example 8.3. The reliability is the probability that the system performance level is lower than the demand w=75, i.e. the system leaves states 4 and 3 the first time. The Markov model can then be constructed as Fig. 8.8.

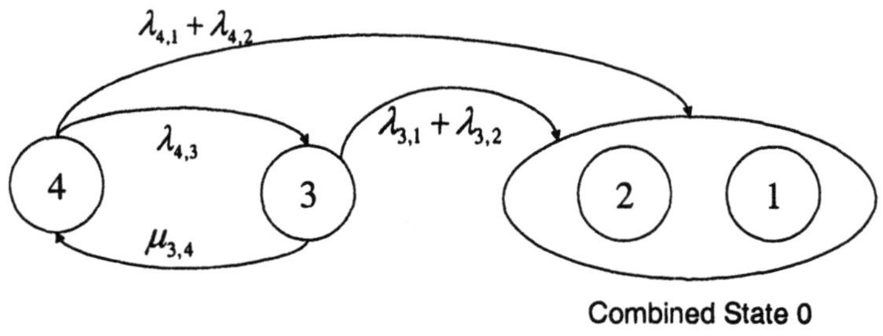

Fig. 8.8. CTMC for the reliability function of Example 8.4.

By substituting the numerical values as given by Example 8.3 into the above equations and solving them, we obtain the state probability functions as

$$P_4(t) = 0.9804\exp(-0.008t) + 0.0196\exp(-1.028t)$$

$$P_3(t) = 0.0196\exp(-0.008t) - 0.0196\exp(-1.028t)$$

$$P_0(t) = 1 - \exp(-0.008t)$$

Then, the reliability function is given by

$$R(t) = P_4(t) + P_3(t) = \exp(-0.008t)$$

■

8.3. A MSS Failure Correlation Model

Most of MSS reliability models assume independence of successive system runs. It is an assumption not valid in reality. This section presents an MSS reliability model based on Markov renewal processes for the modeling of the dependence among successive runs.

8.3.1. Modeling MSS correlated failures

Except the perfect working state, other states in the MSS can be viewed as different types of failure states. Note that if the failures can be of n different types, the total number of possible states for the MSS will be n+1, in which there is a perfect state.

For the correlated MSS with n types of failures and a successful state, a general Markov process can be constructed as follows:

1) Build an n+1-state discrete time Markov chain with transition probability matrix as

$$\mathbf{P}=\begin{bmatrix} P_{00} & P_{01} & \cdots & P_{0n} \\ P_{10} & P_{11} & \cdots & P_{1n} \\ \vdots & \vdots & \vdots & \vdots \\ P_{n0} & P_{n1} & \cdots & P_{nn} \end{bmatrix}$$

2) To overcome the discrete-time property, introduce a process in continuous time by letting the time spent in a transition from state k to state l to have Cdf $F_{k,l}(t)$.

Such a process is attributed to a Semi-Markov Process.

Model for two failure states

When there are two failure states, there will be three states for the MSS after a run; a successful state, Type A failure state and Type B failure state. Type A failure could be a kind of serious failure such as Catastrophic or Critical failure. Type B failure could be less serious than Type A failure such as Minor or Marginal failure.

A common situation is that the system is not able to continue to perform its function when Type A failure occurs, but when Type B failure occurs, the system can still work, although it will have more chances to induce a Type A failure in the next run. The result from a run will affect the probable state in the next run as shown in Fig. 8.9. Here we consider the case when there is no debugging except the resetting or restarting when Type A failure occurs. The transition probability will remain unchanged under this assumption.

Let Z_k be a random variable of the state after a run, and denote by

$$P_{mj} = P\{Z_{k+1} = j \mid Z_k = m\},\ m, j = 0,1,2$$

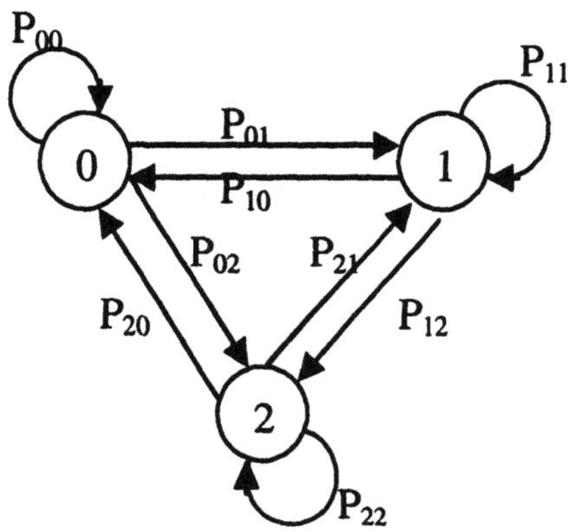

State 0: Successful state after a run
State 1: Type A failure occurs after a run
State 2: Type B failure occurs after a run

Fig. 8.9. Markov interpretation of dependent runs.

The transition matrix is

$$\mathbf{P}=\begin{bmatrix} P_{00} & P_{01} & P_{02} \\ P_{10} & P_{11} & P_{12} \\ P_{20} & P_{21} & P_{22} \end{bmatrix} \tag{8.17}$$

in which

$$\sum_{j=0}^{2} P_{mj} = 1,\ m=0,1,2 \tag{8.18}$$

The unconditional probability of failure on run $(i+1)$ is:

$$P\{Z_{i+1} = j\} = \sum_{m=0}^{2} P_{mj} P\{Z_i = m\}, \; j=0, 1, 2 \tag{8.19}$$

Substituting Eq. (8.18) into the above equation, we have that

$$P\{Z_{i+1} = j\} = (P_{2j} - P_{0j})P\{Z_i = 2\} + (P_{1j} - P_{0j})P\{Z_i = 1\} + P_{0j}, \; j=0,1,2 \tag{8.20}$$

The next step is to develop a model in continuous time, considering the time that the system spends on running. Let $F_{k,l}(t)$ be a Cdf of the time spent in a transition from state k to state l of the DTMC in Fig. 8.9. Here, $F_{k,l}(t)$ is assumed to depend only on the state at the end of each interval in a system run, see e.g. Goseva-Popstojanova & Trivedi (2000) as:

$$F_{0,j}(t) = F_{1,j}(t) = F_{2,j}(t) = F_{\bullet j}(t), \; j=0,1,2$$

With the addition of the $F_{\bullet j}(t)$ to the transitions of discrete time Markov chain, we obtain a Semi-Markov Process as the system reliability model in continuous time.

Model for two failure states with debugging

Furthermore, we assume that after a Type A failure, the system may be debugged and it is an instantaneous fault removing process. Hence, after removing the fault, the transition probability matrix will be changed. When the successive runs are successful or only cause the Type B failure, the system does not have to be debugged and it will continue running in the same way. In this case, the transition probability matrix can then be assumed to be unchanged until a Type A failure happens.

The Markov renewal model is modified as the Fig. 8.10.

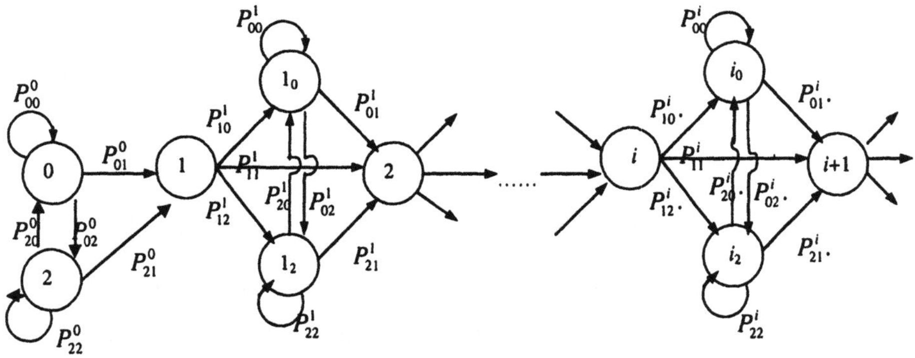

Fig. 8.10. Nonhomogeneous DTMC for system reliability model.

Here '*i*' is the number of Type A failures, which is already detected and removed. During the testing phase, system is subjected to a sequence of runs, making no changes if there is no Type A failure. When a Type A failure occurs on any run, then an attempt is made to fix the underlying fault, which causes the conditional probabilities of the state on the next run to change. The transition probability matrix for the period from the occurrence of the *i*:th Type A failure to the occurrence of the next (*i*+1):st Type A failure, is

$$\begin{bmatrix} P_{00}{}^i & P_{01}{}^i & P_{02}{}^i \\ P_{10}{}^i & P_{11}^i & P_{12}{}^i \\ P_{20}{}^i & P_{21}{}^i & P_{22}{}^i \end{bmatrix} \tag{8.21}$$

Assume S_m is the total number of Type A failures after *m* runs. The sequence S_m provides an alternate description of system reliability model with debugging process considered here. Thus, $\{S_m\}$ defines the DTMC presented in the above Fig. 8.10. All states, i, i_0 and i_2, represent that the Type A failure state has been occupied i times. State i represents the initial state for which $S_m = i$. State i_0 represents all the successful subsequent trials for which $S_m = i$, State i_2 represents all Type B failures subsequent trials for which $S_m = i$.

General model for *n* failure states

The above models can be extended to the case of general multi-state of failures. Assume that the failures can be divided into *n* states, so the MSS totally contains *n*+1 states including the perfect state. Denote again the critical failure type as Type A failure state. When this type of failures occurs, the system will completely stop working and action has to be taken. First we assume there are no changes in the system except resetting and restarting when Type A failure occurs. The transition probability matrix for the successive runs will remained unchanged. The Markov process can be expressed as the Fig. 8.11.

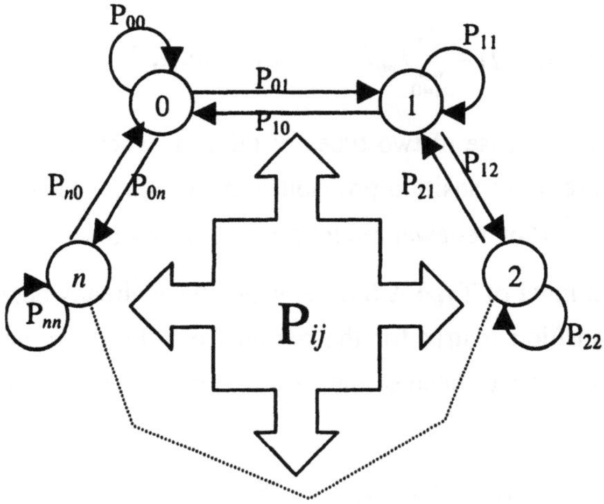

State 0: Successful state after a run

State 1: Type A failure occurs after a run

States 2 to *n*: The other *n*-1 types of failures occur after a run

Fig. 8.11. Markov interpretation for *n*-type correlated state transition.

Denote

$$P_{mj} = P\{Z_{k+1} = j \mid Z_k = m\},\ m, j = 0,1,2,\cdots,n$$

and the transition probability matrix is then

$$\mathbf{P}=\begin{bmatrix} P_{00} & P_{01} & P_{02}\cdots & P_{0n} \\ P_{10} & P_{11} & P_{12}\cdots & P_{1n} \\ P_{20} & P_{21} & P_{22}\cdots & P_{2n} \\ \vdots & \cdots & \cdots & \vdots \\ P_{n0} & P_{n1} & \cdots & \cdots P_{nn} \end{bmatrix}$$

and transition probabilities should satisfy

$$\sum_{j=0}^{n} P_{mj} = 1, \; m=0,1,2,\ldots,n \tag{8.22}$$

The unconditional probability of failure on run (i+1) is:

$$P\{Z_{i+1} = j\} = \sum_{m=0}^{n} P_{mj} P\{Z_i = m\}, \; j=0, 1, 2,\ldots, n \tag{8.23}$$

Similar to the previous case of two types of failures, when there is a debugging after Type A failure, the transition probability matrix changes accordingly. A n-type failure states Markov renewal model can be constructed.

Let 'i' be the number of Type A failure, which have already been detected and removed. The transition matrix for the period from the occurrence of the i:th Type A failure to the occurrence of the next (i+1):st Type A failure, is given as follow:

$$\begin{bmatrix} P_{00}^{\;i} & P_{01}^{\;i} & P_{02}^{\;i}\cdots & P_{0n}^{\;i} \\ P_{10}^{\;i} & P_{11}^{\;i} & P_{12}^{\;i}\cdots & P_{1n}^{\;i} \\ P_{20}^{\;i} & P_{21}^{\;i} & P_{22}^{\;i}\cdots & P_{2n}^{\;i} \\ \vdots & \cdots & \cdots & \vdots \\ P_{n0}^{\;i} & P_{n1}^{\;i} & \cdots & \cdots P_{nn}^{\;i} \end{bmatrix} \tag{8.24}$$

and the transition probability should satisfy

$$\sum_{j=0}^{n} P_{mj}^{\;i} = 1, \; m=0,1,2,\ldots,n$$

Again $\{S_m\}$ defines the DTMC. All the states, i, i_0, i_2,..., i_n, represent that the Type A failure state has been occupied i times. State i represents the first trial for which $S_m = i$. State i_0 represents all the successful subsequent trials for which $S_m = i$, State i_2 to i_n represents Type 2 to n failure states subsequent trials for which $S_m = i$.

8.3.2. Application of the model

The above Markov renewal model can be used to analyze the system performance in both testing phase and validation phase. In testing phase, the system is debugged, so the transition probabilities should change after each Type A failure. However, between two Type A failures, the transition probabilities are constant, so the distribution of time between two successive Type A failures can be easily derived by using the Laplace-Stieltjes transform. The conditional system reliability, which is defined as the survivor time distribution between two Type A failures, can also be obtained.

On the other hand, the probability transition matrix will be constant during the validation phase after the test, because no changes are made to the system during that phase. Hence, the system reliability can be easily calculated.

Some quantitative measures

From a reliability point of view, the time between failures or the number of failures over time is very important. Here, we derive the distribution of the discrete random variable X_{i+1}^j (j=0,2,3...n) defined as the number of runs visiting the j:th state between two successive visits from the i:th Type A failure to the (i+1):st Type A failure.

The probability of every possible number of X_{i+1}^j (j=0,2,3...,n) is given by

$$P\{X_{i+1}^{j} = K_j \mid j = 0,2,3\ldots,n\}$$
$$= \begin{cases} P_{11}^{i} & (\forall K_j = 0|_{j=0,2,3,\ldots,n}) \\ g(K_0, K_2, K_3, \ldots, K_n) & (\exists K_j \neq 0|_{j=0,2,3,\ldots,n}) \end{cases} \quad (8.25)$$

in which $g(K_0, K_2, K_3, \ldots, K_n)$ is the function of $K_0, K_2, K_3, \ldots, K_n$, and K_j denotes the number of runs occupied on the j:th failure state (K_j=0,1,2...). The value of $g(K_0, K_2, K_3, \ldots, K_n)$ can be obtained in principle.

Under the condition of that it visits the j:th state with K_j times (j=0,2,3,...,n) and that Type A failure occurs once between the i:th and (i+1):st Type A failures, the distribution of the time period used for this event can be derived as

$$G(t) = F_{\bullet 0}^{K_0 *}(t) \otimes F_{\bullet 2}^{K_2 *}(t) \otimes F_{\bullet 3}^{K_3 *}(t) \cdots \otimes F_{\bullet n}^{K_n *}(t) \otimes F_{\bullet 1}(t) \quad (8.26)$$

in which $F_{\bullet j}^{K_j *}(t)$ is the K_j-fold convolution of $F_{\bullet j}(t)$ $(j = 0,2,3\ldots,n)$ and K_j can be 0,1,2.... Also, '$\otimes$' denotes the convolution of the two functions.

Denote the distribution of time between the i:th and (i+1):st Type A failures as $F_{i+1}(t)$. Assume T_{i+1} is the random variable of time between the i:th and (i+1):st Type A failure runs. With the above two equations, it can be shown that the distribution of

$$F_{i+1}(t) = P\{T_{i+1} \leq t\} = \sum_{K_0=0}^{\infty} \sum_{K_2=0}^{\infty} \cdots \sum_{K_n=0}^{\infty} P\{X_{i+1}^{j} = K_j \mid j = 0,2,\ldots,n\} \cdot G(t) \quad (8.27)$$

The Laplace-Stieltjes transform of $F_{i+1}(t)$ can be obtained and the inversion of it is straightforward. A closed-form result can be obtained when $F_{\bullet j}(t), (j = 1,2,\ldots n)$ has a rational Laplace-Stieltjes transform.

The reliability of the system after i:th Type A failure is

$$R_{i+1}(t) = 1 - F_{i+1}(t) = P(T_{i+1} > t) \quad (8.28)$$

Some general properties of the inter-failure time can be developed without making other assumptions. For example, the mean time between failures (i and $i+1$ Type A failures) is:

$$E[T_{i+1}] = \int_0^\infty R_{i+1}(t)dt \tag{8.29}$$

or, see e.g. Goseva-Popstojanova & Trivedi (2000)

$$E[T_{i+1}] = -\frac{d\tilde{F}_{i+1}(s)}{ds}\Big|_{s=0} \tag{8.30}$$

Application to the validation phase

After the testing (debugging) phase, the system enters a validation phase to show that it has a high reliability prior to actual use. In this phase, no changes are made to the system. Here, we use the two-type failure case as an illustration. Similar procedures can be implemented in solving general n-type failure problems.

First we consider the independent condition, that is,

$$P_{0j} = P_{1j} = P_{2j} = P_{\bullet j},\ j\text{=}0,1,2$$

If the state is not a Type A failure after a run, the system is reliable until the Type A failure occurs. The reliability in a run is $1 - P_{\bullet 1}$. The reliability for m successive runs is defined as the probability that m successive independent test runs are conducted without Type A failure, which can be derived as:

$$R(m) = (1 - P_{\bullet 1})^m = (P_{\bullet 0} + P_{\bullet 2})^m \tag{8.31}$$

Given a confidence level α, if $R(m) \geq \alpha$, we can say that the system is reliable in successive m runs without Type A failure with α confidence. In order to satisfy this condition, the value of $P_{\bullet 1}$ should satisfy

$$(1 - P_{\bullet 1})^m \geq \alpha \tag{8.32}$$

Given a confidence level α, we can obtain an upper confidence bound on $P_{\bullet 1}$, which is denoted by P_1^*. Solving $(1-P_{\bullet 1}^*)^m=\alpha$, we obtain the upper bound

$$P_{\bullet 1}^*=1-\alpha^{1/m} \tag{8.33}$$

This can help to test whether the system can be certified or not, i.e., if $P_{\bullet 1}\le P_{\bullet 1}^*$, the system is certified with α confidence to say that the system is reliable in n successive runs without Type A failure.

Now consider a sequence of possibly dependent system runs. During the validation phase, the system is not changing, i.e., P_{ij} does not change. That is, the sequence of runs can be described by the homogeneous DTMC with the transition probability matrix. Assume that the DTMC is steady, i.e., each run has the same failure-probability:

$$P\{Z_{i+1}=j\}=P\{Z_i=j\}=\sum_{m=0}^{n}P_{mj}P\{Z_i=m\},\ j\text{=0,1,2} \tag{8.34}$$

Let $P_j=P\{Z_i=j\}$ and substitute it into the above equation to get

$$P_j=\sum_{m=0}^{n}P_{mj}P_m,\ j\text{=0,1,2} \tag{8.35}$$

Solve the above equations to obtain unconditional probability of failure on run as

$$P_2=\frac{P_{01}P_{12}+P_{02}-P_{11}P_{02}}{(1-P_{11}+P_{01})(1-P_{22}+P_{02})-(P_{12}-P_{02})(P_{21}-P_{01})} \tag{8.36}$$

$$P_1=\frac{P_{02}P_{21}+P_{01}-P_{22}P_{01}}{(1-P_{22}+P_{02})(1-P_{11}+P_{01})-(P_{21}-P_{01})(P_{12}-P_{02})} \tag{8.37}$$

$$P_0=1-P_1-P_2 \tag{8.38}$$

The reliability for m successive runs will be

$$R(m)=(1-P_1)^m=(P_0+P_2)^m \tag{8.39}$$

An example is given here to illustrate the procedure.

Example 8.5. Suppose the distribution of the execution time of each run is exponential so that

$$F_{\bullet j}(t) = 1 - \exp(-\mu_j t), j=0,1,2$$

Let $\mu_0 = 0.302, \mu_1 = 0.3, \mu_2 = 0.297$ as illustration. In the operational phase we can estimate the transition probability matrix from empirical data of successive runs. The following transition probability matrix is used as illustration

$$\mathbf{P}=\begin{bmatrix} P_{00} & P_{01} & P_{02} \\ P_{10} & P_{11} & P_{12} \\ P_{20} & P_{21} & P_{22} \end{bmatrix} = \begin{bmatrix} 0.7 & 0.1 & 0.2 \\ 0.3 & 0.1 & 0.6 \\ 0.1 & 0.2 & 0.7 \end{bmatrix}$$

Substitute those values into Eq. (8.27), we can obtain the Laplace-Stieltjes transform equation and then invert it to get the Cdf of the time between failures as:

$$F(t) = 1 - 0.053e^{-0.3t} - 0.038e^{-0.302t} - 0.051e^{-0.297t}$$

$$-0.041e^{-0.55t} - 0.45e^{-0.05t} - 0.045e^{-0.51t} - 0.054e^{-0.342t}$$

$$-0.157e^{-0.259t} - 0.127te^{-0.09 \cdot t} - 0.111e^{-0.09t}$$

This equation implies that when successive runs are dependent, the Cdf of the time between failures is a mixture of exponential distributions. Fig. 8.12 displays the distribution of $F(t)$.

Using the distribution function, the mean time to failure can be obtained as

$$E[T] = \int_0^{\infty} R_{i+1}(t)dt = 27.31 \text{ (hours)}$$

The unconditional probability of the three different states can be calculated through Eqs. (8.36-8.38)

$$P_2 = \frac{P_{01}P_{12} + P_{02} - P_{11}P_{02}}{(1 - P_{11} + P_{01})(1 - P_{22} + P_{02}) - (P_{12} - P_{02})(P_{21} - P_{01})} = 0.522$$

$$P_1 = \frac{P_{02}P_{21} + P_{01} - P_{22}P_{01}}{(1 - P_{22} + P_{02})(1 - P_{11} + P_{01}) - (P_{21} - P_{01})(P_{12} - P_{02})} = 0.152$$

$$P_0 = 1 - P_1 - P_2 = 0.326$$

The steady probability for the system to be reliable is

$$P_0 + P_2 = 0.326 + 0.522 = 0.848$$

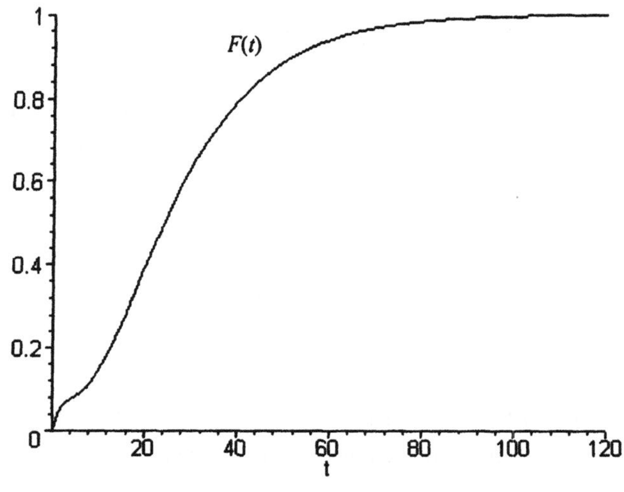

Fig. 8.12. Cdf of the time between failures.

■

8.4. Notes and References

For the multi-state systems, the book of Lisnianski & Levitin (2003) summarized many MSS reliability models, which can provide the readers a complement view to this chapter. They have carried extensive research on this topic. The book describes many MSS reliability models of different structures including series, parallel, bridge and distributed networks etc, and under different environments including weighted voting systems, consecutively connected systems, sliding window systems and so on.

Xue & Yang (1997) showed that multi-state reliability dynamic analysis could be transformed to a set of 2-state ones by using some generalized reliability parameters. Bukowski & Goble (2001) studied the MTTF of the MSS. Kolowrocki (2001) studied the MSS with components having exponential reliability functions with different transition rates between subsets of their states, which introducing the aging concept into the components of the MSS. Levitin & Lisnianski (2001) considered vulnerable systems, which could have different states corresponding to different combinations of available elements composing the system. In real systems, a multilevel protection is often used, for example, in defense-in-depth design methodology (Fleming & Silady, 2002). The multilevel protection means that a subsystem and its inner level protection are in their turn protected by the protection of the outer level, which has been studied by Levitin (2003). Yeh (2003) presented an interesting model for the network reliability by assuming the nodes and links composing the network are of multiple states.

Levitin & Lisnianski (2003) formulated the optimization problem of designing structure of series-parallel multi-state system (including choice of system elements, their separation and protection) in order to achieve a desired level of system survivability by the minimal cost. Liu *et al.* (2003) also presented a neural network to solve this optimization problem. Recently, Levitin *et al.* (2003) further extended it to include multiple levels of protections and presented a multi-processor genetic algorithm to solve it.

CHAPTER 9

OPTIMAL SYSTEM DESIGN AND RESOURCE ALLOCATION

In the design of computing systems, some important decision problems need be solved. These problems could be the determination of optimal number of distributed hosts, the system structure and the network architecture. The objectives could be to maximize the reliability, to minimize the cost, or both.

Besides the optimal system design, the problem of optimally allocating limited resources (such as time, manpower, programs or files) on the computing systems are also of great concern. Given limited resources, different allocation strategies will cause different system reliability and cost. In order to make the best of the resources, their allocations must be carefully considered.

This chapter discusses some of these optimization problems. The optimal number of redundant hosts for a distributed system design is first presented. Optimal testing resource allocation problems on either independent modules or dependent versions of software are discussed. Finally, the optimization of grid architecture design and the grid service integration problems are studied.

9.1. Optimal Number of Hosts

An important goal in computing system design is to achieve a high reliability or availability through some kind of redundancy (such as redundant hosts) or fault tolerance. Many systems are developed in the environment with redundant hosts. The number of hosts has significant influence on the cost and system availability because it can be very costly while they are able to improve system availability easily. The objective here is to minimize the total cost based on the following cost model.

9.1.1. The cost model

To illustrate the relationships among the decisions and cost, an influence diagram which provides simple graphical representations of decision situations, is displayed in Fig. 9.1. Different decision elements are shown in the influence diagram as of different shapes, see e.g., Clemen (1995 pp. 50-65).

The number of redundant hosts affects the optimal decision of the release time. Both the number of redundant hosts and release time affect the system availability. These factors determine the development cost. The number of hosts also determines the cost of redundant hosts. The release time determines the rewards or penalty depending on whether the release is before or after the deadline. If the system is unavailable after release, a risk cost is incurred. Hence, the cost of redundant hosts, the development cost, reward and penalty should be considered together when deriving the total expected cost. Each cost component will be described in the following.

Cost of redundant hosts

The cost function for a multi-version fault-tolerant system can be described as a linear function to the number of versions as

$$C_h(N) = a_1 N + b_1 \tag{9.1}$$

where N is the number of hosts, b_1 is a constant, and a_1 is defined as the expected cost per host. Here we have assumed the redundant hosts used in the system are of the same type.

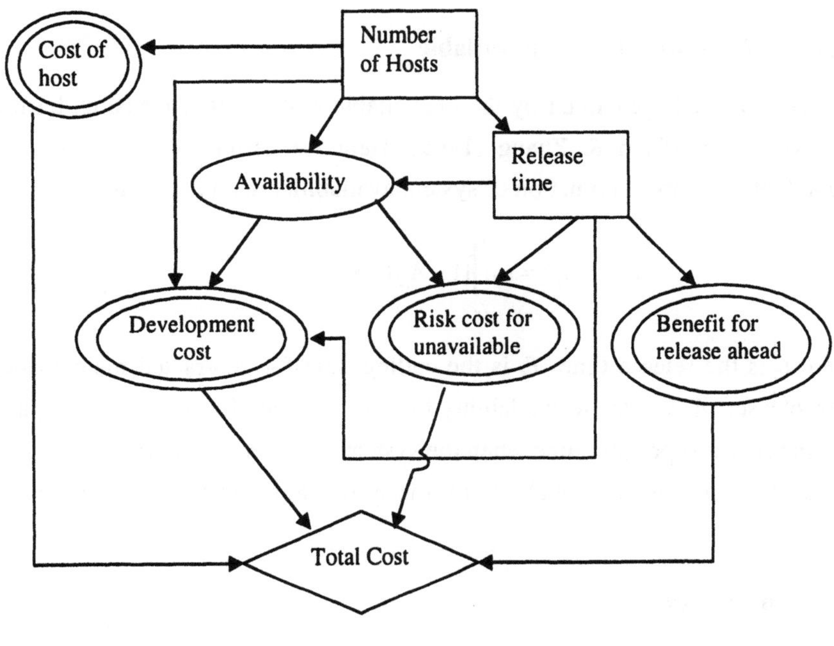

Fig. 9.1. Influence diagram for the cost affected by redundant hosts.

Reward for early release

Usually there is a deadline for release. This is the case when the penalty cost for delay is very high. On the other hand, there is a reward for releasing the system earlier. We assume b_2 is a constant rewarded if the system can be released in time, no matter how early the release time is and a_2 is the expected reward per unit time before the deadline. The reward function of the release time can be expressed as

$$B(t_r) = a_2(T_d - t_r) + b_2, \quad t_r \leq T_d \tag{9.2}$$

where T_d is the deadline for release, t_r is the release time so that $T_d - t_r$ is the time ahead of the schedule.

Risk cost for system being unavailable

This cost factor is generated by the unavailable system after releasing, termed risk cost as in Pham & Zhang (1999). Here we assume the risk cost for unavailable system is a function of system availability and release time:

$$C_r(N,t_r) = a_3 \int_{t_r}^{T_e} [1 - A_N(t)]dt \tag{9.3}$$

where t_r is the release time, T_e is the ending time for contracted maintenance after release, $A_N(t)$ is the availability function at time t for N-host system, and a_3 is the risk cost per unit time when the system is not available. In the equation above, $1 - A_N(t)$ is the probability for the system to be unavailable at time t.

Development cost

The development cost function for a single software module proposed in Kumar & Malik (1991) is

$$C_i(R_i) = H_i \exp(B_i R_i - D_i) \tag{9.4}$$

where H_i, B_i and D_i are constants and R_i is the individual module software reliability achieved at the end of testing.

Then, the total expected cost can be expressed as

$$C(N,t_r) = C_h(N) + C_r(N,t_r) + C_t(R(t_r)) - B(t_r) \tag{9.5}$$

9.1.2. System availability

The system availability model for a homogeneous distributed software/hardware system can be obtained straightforward from Chapter 6.3. A numerical example is shown below.

Example 9.1. Suppose $K_0 = 32$ and λ =0.006, λ_h =0.01, μ_h =0.1 and μ_s =0.13, the system availability for different number of hosts can be obtained from the analysis presented in Chapter 6.3. The results are depicted in Fig. 9.2.

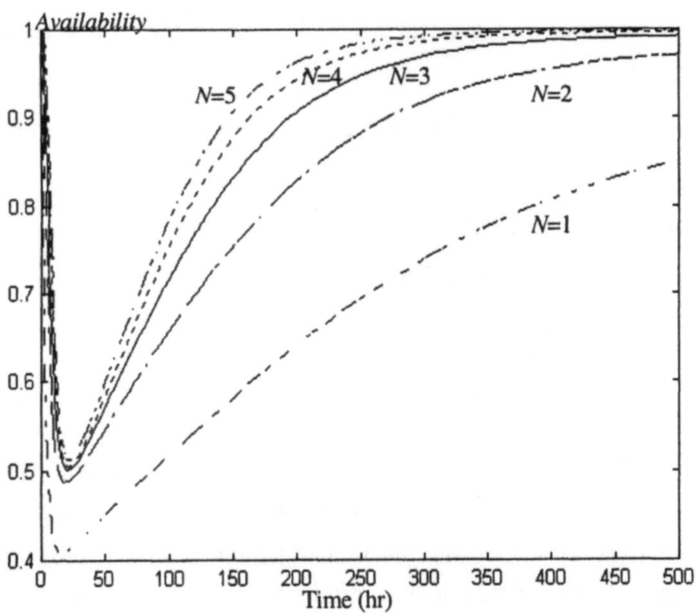

Fig. 9.2. System availability for different number of redundant hosts.

■

We can observe that when the number of redundant hosts increases, the system availability increases. The system availability function can be used in the optimization model which will be described in the following.

9.1.3. Optimization model and solution procedure

The optimization model is based on the cost criteria and the decision variables are the number of redundant hosts and the release time. Its objective is to minimize the expected total cost. There are three types of constraints in this decision problem. First, the customers may require a least system availability A^* after the release. Second, there is a deadline for the system to be released so the release time should be earlier than that. Finally, the customers may limit the maximum number of redundant hosts N^* due to their budget and other physical restrictions.

That is, the decision variables are N and t_r, and the optimization model is to

Minimize: $$C(N,t_r) \tag{9.6}$$

Subject to: $$A_N(t_r) \geq A^* \geq 0 \tag{9.7}$$

$$0 \leq t_r \leq T_d \tag{9.8}$$

$$N = 1,2,3,\ldots,N^* \tag{9.9}$$

where A^* is the required system availability after the release, T_d is the deadline for release and N^* is the maximum number of redundant hosts allowed. If there is no such constraint, we can assume a large enough value of N^* in this model. However, usually only a small number of redundant hosts will be practical.

To obtain an optimal solution, the solving procedures are described as follows:

Step 1: Obtain the system availability function of the distributed system with N redundant hosts.

Step 2: Derive each cost function and obtain the expected total cost.

Step 3: Let N take each integer value from 1 to N^* to obtain the expected total cost and save the results from $C(1,t_r)$ to $C(N^*,t_r)$.

Step 4: For each $C(1,t_r)$ to $C(N^*,t_r)$, compute the optimal release time, and save the results as OpTr(1) to OpTr(N^*), so that the minimum expected total cost is obtained and saved in MinC(1) to MinC(N^*).

Step 5: Compare the minimum total expected cost from MinC(1) to MinC(N^*)to select the optimum number of redundant hosts $OpN = Min(\text{MinC(n)})$ (n=1,2,..., N^*).

The above procedure can be easily realized in computer programs. A numerical example is presented to illustrate the optimization procedures.

Example 9.2. Company X is awarded a contract to develop a telephone switching system. In this case, the hardware hosts are brought from external suppliers, but the software is developed in house and tested with the system. The main question is how many redundant hosts are needed and also we are interested in when the system can be released so that the total cost is minimized. For illustrative purpose, the following input values are used:

1) The system availability needs to be higher than 0.88 when it is released.

2) The deadline for releasing the system is 800 hours from now.

3) The penalty cost for unavailable system is about $8000 per hour during the first 300 hours after release.

4) Each host costs $17600 and a fixed fee for all the hosts is $1293.

5) The maximum number of redundant hosts is five.

6) If the company can release the system earlier than the deadline, there is a constant reward of \$2123.7 and a variable reward of \$31.5 per hour.

Based on the conditions and the assumptions given above, the values of the parameters can be obtained as

$$a_1 = 17600, \; b_1 = 1293, \; a_2 = 31.5,$$

$$b_2 = 2123.7, \; a_3 = 8000, \; T_d = 800 \text{ hours},$$

and $T_e = 800 + 300 = 1100$ hours.

The parameters for software development cost are assumed as H=10232, B=16, D=14. The optimization problem can be solved with the required system availability when releasing, A^*, of 0.88 and the maximum number of redundant hosts, N^*, equal to 5.

Here we assume the system is a kind of homogeneous distributed software/hardware system whose availability function is depicted by Fig. 9.2. With the values of parameters given above, we can obtain the total mean cost through Eq. (9.5) as

$$C(N,t_r) = 17600N + 31.5t_r + 8000 \int_{t_r}^{1100} [1 - A_N(t)]dt$$

$$+10233\exp\{16R(t_r) - 14\} - 26030.7$$

Finally, the total expected cost as a function of release time for different number of redundant hosts are depicted by Fig. 9.3 and the overall results are given in Table 9.1.

Table 9.1. Numerical values of the minimum cost for different *N*.

N	1	2	3	4	5
MinC(N)	326970	153060	116690	104580	110880
OpTr(N)	800	800	324.2	261.7	232

From Table 9.1, the global minimum cost is 104580 (Units) with the number of redundant hosts N=4 and the optimum release time t_r =261.7 (hrs). The optimum results indicate that there should be four redundant hosts and the system is tested for 261.7 hours.

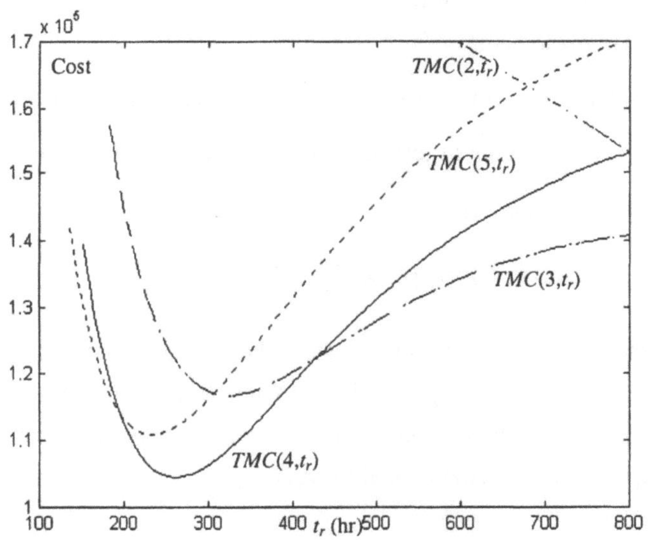

Fig. 9.3. Total expected cost *vs.* release time of different number of hosts.

■

9.2. *Resource Allocation - Independent Modules*

Testing-resource refers to the resource expenditures spent on software testing, e.g., man-power and available time, etc. During the testing stage, a project manager often faces various decision-making problems such as how to allocate available time (the time before deadline) among the modules and how to assign personnel, etc. In order to combine these two types of resources (man-power and available time), we define a term called total testing time that is calculated by multiplying the number of personnel with the available time. Each unit of the total testing time represents the resource of one person to work for one unit of time. Here the testing-resource is referred to as total testing time and we use the term testing-resource as an exchangeable one with the term total testing time.

For the optimal testing-resource allocation problem, the following assumptions are made here:

(a) n modules in a software are independent during the unit-testing phase.

(b) After T_i unit time of testing, the failure rate of module i is $\lambda_i(T_i)$.

The reliability of module i is

$$R_i(x \mid T_i) = \exp\{-\lambda_i(T_i)x\}, \quad x \geq 0 \tag{9.10}$$

where x is the operational time after testing. Note that in the above, we have used the operational reliability definition (Yang & Xie, 2000) as it is more common that after the release, there will be no reliability growth, and hence the failure rate will remain constant equal to $\lambda_i(T_i)$.

9.2.1. Allocation on serial modular software

If the software system fails whenever there is a failure with any of the modules that the software system is composed of, then it is called a serial software system. For many modular software systems this is right the case. The structure for such a system is illustrated in Fig. 9.4.

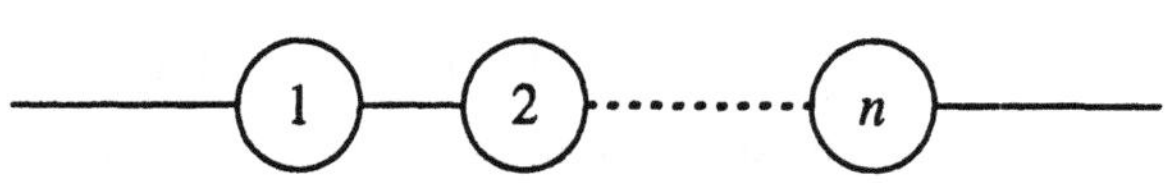

Fig. 9.4. Structure of a serial software system.

Denote by T_i the testing-time allocated to module i. After T_i unit of time of testing, the failure rate of module i is $\lambda_i(T_i)$. The reliability of module i is

$$R_i(x \mid T_i) = \exp\{-\lambda_i(T_i)x\}, \quad x \geq 0, \; i = 1, 2, \ldots, n$$

The reliability of the whole software system is given by

$$R(x \mid T_1, \cdots T_n) = \prod_{i=1}^{n} R_i(x \mid T_i) = \exp\left\{-x \sum_{i=1}^{n} \lambda_i(T_i)\right\} \tag{9.11}$$

The optimal testing-resource allocation problem is formulated as

$$\textbf{Maximize} \quad R(x \mid T_1, \cdots T_n) = \exp\left\{-x \sum_{i=1}^{n} \lambda_i(T_i)\right\} \tag{9.12}$$

$$\textbf{Subject to} \quad \sum_{i=1}^{n} T_i \leq T \tag{9.13}$$

$$T_i \geq 0, \quad i = 1, 2, \ldots, n \tag{9.14}$$

The formulation above is equivalent to minimizing the sum of failure occurrence rates, i.e.,

$$\textbf{Minimize} \quad \sum_{i=1}^{n} \lambda_i(T_i) \tag{9.15}$$

It can be noted that the general formulation presented above does not require a particular model for the mean value function and thus it has much flexibility. In fact, we could even use different software reliability models for different modules.

In order to obtain a general solution to this problem, the Lagrangian is constructed as

$$L = \sum_{i=1}^{n} \lambda_i(T_i) + \lambda\left(\sum_{i=1}^{n} T_i - T\right) \tag{9.16}$$

The necessary and sufficient conditions for the minimum are (Bazaraa *et al.*, 1979, p. 149)

$$\frac{\partial L}{\partial T_i} = \frac{\partial \lambda_i(T_i)}{\partial T_i} + \lambda \geq 0, \qquad i = 1, 2, \cdots, n \tag{9.17}$$

$$T_i \frac{\partial L}{\partial T_i} = 0, \quad i = 1, 2, ..., n \tag{9.18}$$

$$\lambda\left(\sum_{i=1}^{n} T_i - T\right) = 0 \tag{9.19}$$

$$\lambda \geq 0 \tag{9.20}$$

The optimal solution $T_1^*, T_2^*, ..., T_n^*$ can be obtained by solving the above equations numerically. Define $g_i(t) \equiv -d\lambda_i(t)/dt$, then an equivalence of Eq. (9.17) is

$$\lambda \geq g_i(T_i) \tag{9.21}$$

For most software reliability models, $g_i(t)$ is a positive and non-increasing function. It is shown in Yang & Xie (2001) that if $g_i(t) > 0$ and $g_i(t)$ is non-increasing on $t \in [0, T]$, let

$$A_i \equiv g_i(0), \quad i = 1, 2, ..., n \tag{9.22}$$

Then, if we reorder software modules 1, 2, ..., n such that $A_1 \geq A_2 \cdots \geq A_n$, the optimal solution to the testing-resource allocation problem is:

$$T_i^* = \begin{cases} g_i^{-1}(\lambda^*) & i = 1,\ 2,\ ..., k \\ 0 & i = k+1, ..., n \end{cases} \tag{9.23}$$

where λ^* satisfies $\sum_{i=1}^{k} g_i^{-1}(\lambda^*) = T$, and k satisfies $A_k > \lambda^* \geq A_{k+1}$.

From the results above, the optimum solution can be obtained by the following iteration algorithm.

Step 1. Compute A_i using Eq. (9.22).

Step 2. Set $l = 1$.

Step 3. Obtain λ_l by solving the following equation:

$$\sum_{i=1}^{l} g_i^{-1}(\lambda_l) = T \tag{9.24}$$

Step 4. If $A_l > \lambda_l \geq A_{l+1}$, then $\lambda^* = \lambda_l$ and the optimal solution can be obtained by Eq. (9.23), then stop. Otherwise set $l = l + 1$ and go back to Step 3.

Example 9.3. Assume that the software system is composed of three modules for which the testing processes follow the logarithmic Poisson execution time model (Musa & Okumoto, 1984). That is,

$$m_i(t) = \frac{\ln(\alpha_i \varphi_i t + 1)}{\varphi_i}, \quad i = 1,2,3$$

It can be shown that

$$g_i(t) = \frac{\alpha_i^2 \varphi_i}{(\alpha_i \varphi_i t + 1)^2}, \quad i = 1,2,3$$

are positive and strictly decreasing on $t \in [0, \infty)$. The optimization algorithm described in previous section can be used. In this case, $A_i = \alpha_i^2 \varphi_i$. The solution to Eq. (9.24) is

$$\lambda_l = \left(\frac{\sum_{i=1}^{l} \frac{1}{\sqrt{\varphi_i}}}{T + \sum_{i=1}^{l} \frac{1}{\alpha_i \varphi_i}} \right)^2$$

and Eq. (9.23) becomes

$$T_i^* = \begin{cases} \dfrac{1}{\sqrt{\varphi_i \lambda^*}} - \dfrac{1}{\alpha_i \varphi_i} & i = 1, \cdots, k \\ 0 & i = k+1, \cdots, 3 \end{cases}$$

Suppose that the parameters of the three modules have been estimated by historical testing data and are summarized in Table 9.2, and an additional 5000 CPU hours of testing-time is available to be allocated among these three modules. By solving the optimization problem as described in previous section, the optimal allocation is obtained and shown in Table 9.2.

Table 9.2. Estimated parameters and optimal allocation for Example 9.3.

Module	α_i	φ_i	T_i^*
1	16.0	0.09	1534
2	6.0	0.03	2653
3	1.5	0.32	813

The reliability of the software system after the additional 5000 hours of testing is:

$$R(x \mid T_1 = 1534, T_2 = 2653, T_3 = 813) = \exp(-0.02361x), \quad x \geq 0$$

■

9.2.2. Allocation on parallel modular software

The system is assumed to be a parallel redundant system (Fig. 9.5). For such a software system, the system will fail only when all modules fail. The achieved reliability of the system after unit testing phase is

$$R(x \mid T_1, \cdots, T_n) = 1 - \prod_{i=1}^{n} [1 - R_i(x \mid T_i)], \quad x \geq 0 \tag{9.25}$$

where $R_i(x \mid T_i)$ is the operational reliability of module *i*.

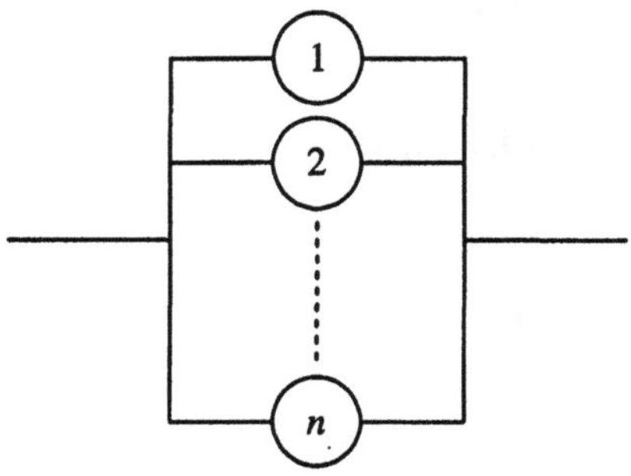

Fig. 9.5. Structure of a parallel redundant software system.

The optimal testing-resource allocation problem is formulated as

$$\textbf{Maximize} \quad R(x \mid T_1, \cdots, T_n) = 1 - \prod_{i=1}^{n} [1 - \exp\{-\lambda_i(T_i)x\}] \tag{9.26}$$

$$\textbf{Subject to} \quad \sum_{i=1}^{n} T_i \le T \tag{9.27}$$

$$T_i \ge 0, \quad i = 1, 2, \ldots, n \tag{9.28}$$

An equivalence of Eq. (9.26) is:

$$\textbf{Minimize} \quad \sum_{i=1}^{n} \ln[1 - \exp\{-\lambda_i(T_i)x\}] \tag{9.29}$$

Now the optimal testing resource allocation problem is formulated by the above equations. The Lagrangian is constructed as:

$$L = \sum_{i=1}^{n} \ln[1 - \exp\{-\lambda_i(T_i)x\}] + \lambda\left(\sum_{i=1}^{n} T_i - T\right)$$

The necessary and sufficient conditions for the minimum are

$$\frac{\partial L}{\partial T_i} = \frac{x}{\exp\{\lambda_i(T_i)x\}-1}\frac{\partial \lambda_i(T_i)}{\partial T_i} + \lambda \geq 0\,,\ i = 1, 2, \cdots, n \tag{9.30}$$

$$T_i \frac{\partial L}{\partial T_i} = 0\,, \quad i = 1, 2, \ldots, n \tag{9.31}$$

$$\lambda\left(\sum_{i=1}^{n} T_i - T\right) = 0 \tag{9.32}$$

$$\lambda \geq 0 \tag{9.33}$$

The optimal solution $T_1^*, T_2^*, \ldots, T_n^*$ can be obtained by solving the above equations numerically.

9.2.3. Allocation on mixed parallel-series modules

The Fig. 9.6 is the structure of a mixed parallel-series modular software system. There are *n* groups of parallel modules and *m* serial modules.

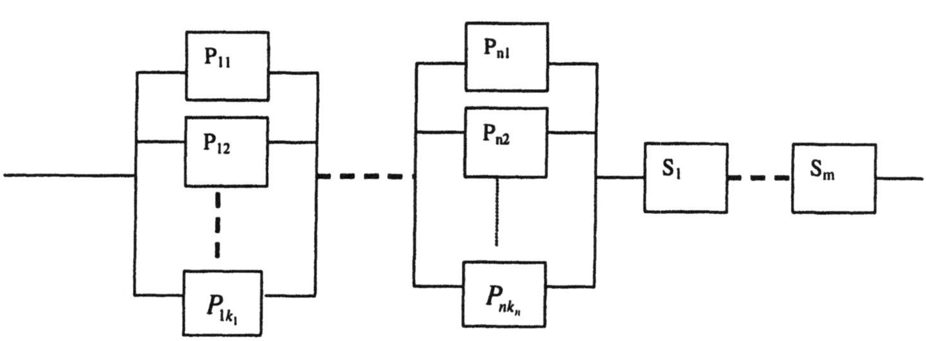

Fig. 9.6. The structure of a parallel-series modular software system.

Single objective of maximizing reliability

The reliability for this parallel-series modular software system is calculated as following equation

$$R(x \mid T) = \prod_{l=1}^{n} \left\{ 1 - \prod_{i=1}^{k_l} [1 - R_{li}(x \mid T_{li})] \right\} \prod_{j=1}^{m} R_j(x \mid T_j) \tag{9.34}$$

where T_i is the testing time allocated to module i. Then, the following optimization model is to maximize system reliability:

$$\textbf{Maximize } R(x \mid T_{li}, T_j) = \prod_{l=1}^{n} \left\{ 1 - \prod_{i=1}^{k_l} [1 - R_{li}(x \mid T_{li})] \right\} \prod_{j=1}^{m} R_j(x \mid T_j) \tag{9.35}$$

$$\textbf{Subject to } \sum_{l=1}^{n} \sum_{i=1}^{k_l} T_{li} + \sum_{j=1}^{m} T_j \le T \tag{9.36}$$

$$T_{li}, T_j \ge 0$$

in which T is the total resource of time consuming in all modules of parallel group (T_{li}) and serial modules (T_j).

Multiple objectives of maximizing reliability and minimizing cost

Assume that the cost function of Module i is $C_i(R_i)$ in which R_i is the reliability for the i:th module. The total cost in the parallel-series modular software system of Fig. 9.6 will be

$$C(R_{li}, R_j) = \sum_{l=1}^{n} \sum_{i=1}^{k_l} C_{li}(R_{li}) + \sum_{j=1}^{m} C_j(R_j) \tag{9.37}$$

where

$\sum_{i=1}^{k_l} C_{li}(R_{li})$ is the total cost of the l:th groups of parallel modules

$\sum_{l=1}^{n}\sum_{i=1}^{k_l} C_{li}(R_{li})$ is the total cost of all the n groups of parallel modules, and

$\sum_{j=1}^{m} C_j(R_j)$ is the total cost of all the series modules.

Here, we adopt the cost function for individual module i shown by Eq. (9.4), proposed in Kumar & Malik (1991).

The optimal testing-resource allocation problem can then be formulated with two objectives as

$$\text{1) } \textbf{Maximize} \quad R(x \mid T_{li}, T_j) = \prod_{l=1}^{n}\left\{1 - \prod_{i=1}^{k_l}[1 - R_{li}(x \mid T_{li})]\right\}\prod_{j=1}^{m} R_j(x \mid T_j) \tag{9.38}$$

$$\text{2) } \textbf{Minimize} \quad C(R_{li}, R_j) = \sum_{l=1}^{n}\sum_{i=1}^{k_l} C_{li}(R_{li}) + \sum_{j=1}^{m} C_j(R_j) \tag{9.39}$$

$$\textbf{Subject to} \quad \sum_{l=1}^{n}\sum_{i=1}^{k_l} T_{li} + \sum_{j=1}^{m} T_j \le T \tag{9.40}$$

$$T_{li}, T_j \ge 0$$

in which T is the total resource of time consuming in every modules of parallel group (T_{li}) and serial modules (T_j).

For mixed parallel-series modular software, it is difficult to solve them, so the heuristic algorithms such as genetic algorithm, simulation annealing or Tabu search can be applied. Dai *et al.* (2003b) presented a genetic algorithm to solve the above multi-objective allocation problems. Here an example of this type is illustrated with that genetic algorithm.

Example 9.4. The structure of this 8 modules example is shown in Fig. 9.7.

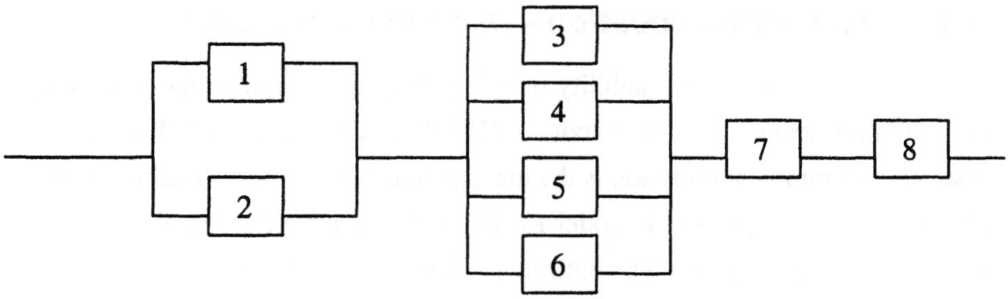

Fig. 9.7. The structure of a complex parallel-series modular system.

We use the GO-model for illustration. The mean value function is:

$$m_i(t) = a_i[1 - \exp(-b_i t)], \quad i = 1,2,...8 \tag{9.41}$$

We assume here that the total testing time is 23000 hours and x is 200 hours to complete the given task. The values of parameters and optimal solution out of the genetic algorithm are given in the following Table 9.3 where T_i^* (i=1,2,...,8) is the optimal allocated testing time on different modules.

Table 9.3. The parameters of parallel-series modular software system.

Modules	a_i	b_i	H_i	B_i	D_i	T_i^*
1	210	0.00051	3.493	6.011	4.97	93.47
2	199	0.00059	3.503	6.12	4.93	10522
3	453	0.00048	3.498	6.012	4.995	0
4	345	0.00058	3.498	6.001	4.997	54.11
5	258	0.00063	3.499	6.002	4.995	60.48
6	221	0.00074	3.5015	6.15	4.97	8822.8
7	33.99	0.00579	3.495	6.01	4.98	2190.83
8	32.32	0.00593	3.500	6.005	4.01	1256.31

■

9.3. Resource Allocation - Dependent Modules

A method to increase the reliability of safety critical software is the *N*-version programming technique, e.g. Avizienis (1985). In the analysis of this type of systems, a common assumption is the independence of different versions. In the following, we first present a model for the dependent *N* versions of software. Then, based on the model, optimum allocation problem of the testing resource/time on the dependent *N* versions is discussed.

9.3.1. Reliability analysis for dependent *N*-version programming

The *N*-version programming involves the execution of multiple versions of software. A voting scheme matches and tests the outputs, and then determines a final result. There are various voting schemes. Here we use the voting scheme of "selecting the first qualified result", which is explained in details in Belli & Jedrzejowicz (1991). In this voting scheme, if any one version among the *N* versions of software passes a test, the voter will select it as the final result no matter whether the other versions are qualified or not.

Decomposition by multi-component modeling

In the *N*-version software, any *j* versions may fail at the same time because of certain common cause failures. For example, if *j* versions of the *N*-version software share a common subroutine, these *j* versions may fail simultaneously. We define a parameter for such failure, called *dependence level*, by the number of simultaneously failed versions caused by the failure.

We denote $M_{j,k}$ as the "components" that correspond to different common cause failures, where j (j=1,2,...,N) is the dependent level that correlates any j out of N versions and k (k=1,2,..., $K_{N,j}$) represents the k:th component among all the j:th dependent components ($M_{j\bullet}$), where

$$K_{N,j} = \binom{N}{j}$$

If all those failures with the *j*:th dependent level are numbered by *k* (k=1,2,..., $K_{N,j}$), $M_{j,k}$ can represent all the failures with different dependent levels, respectively. The total number of all "components" $M_{j,k}$ (j=1,2,...,N; k=1,2,..., $K_{N,j}$) is equal to $2^N - 1$.

The *N* dependent versions of software can be decomposed into the mutually exclusive $2^N - 1$ components. Note that the *N* versions may not be physically separated. An example of three-version programming is illustrated below.

Example 9.5. Consider a fault-tolerant system with three versions of software which might be dependent. The three dependent versions are correlated as shown in Fig. 9.8 and the states can be decomposed into 7 mutually exclusive parts, called components here.

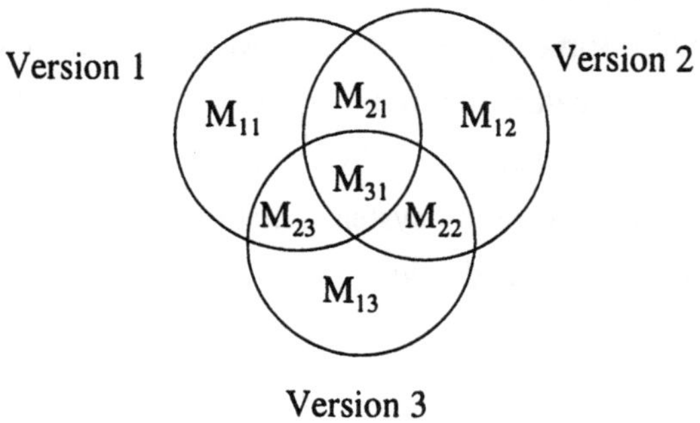

Fig. 9.8. Three dependent versions of software.

Let $M_{1,k}$ (k=1,2,3) denote the failures that affect only the k:th version without influence on the other two versions; $M_{2,k}$ (k=1,2) denote the common cause failures that correlate the k:th and (k+1):st versions without influence on the other one version; $M_{2,3}$ represents the failure that correlates the first and the third versions; and $M_{3,1}$ denotes the failures that correlate all the three versions.

The reliability block diagram for those components can be built as shown in Fig. 9.9.

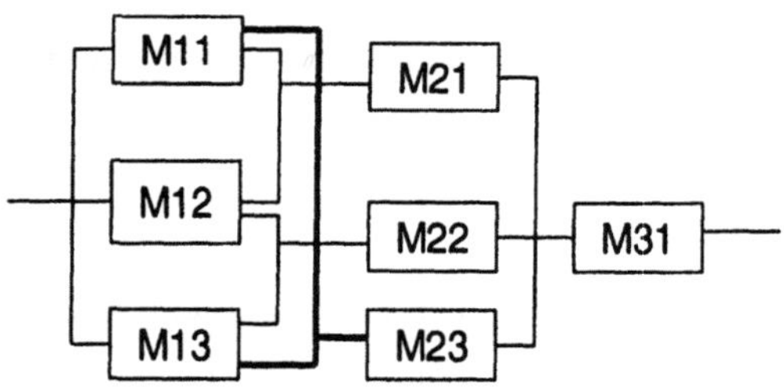

Fig. 9.9. Reliability block diagram of the decomposed components.

The reliability block diagram is complex containing not only many parallel-series units but also some bridge structures. Moreover, the diagram will become much more complicated for four or more versions. Hence, the reliability estimation for dependent N-version programming is not straightforward. In order to analyze the system reliability based on our above model, a general approach is presented below.

System reliability function

The reliability of a component $M_{j,k}$ is defined as the probability for the corresponding common cause failure not to occur, which is denoted by

$R_{j,k}(t)$. The software reliability of the dependent N-version programming is defined as the probability that at least one version of software can achieve the task successfully. The software reliability function at time t can be expressed as

$$R(t) = \Pr(\text{at least one version of software is reliable at time } t) \qquad (9.42)$$

Let $E_i(t)$ represent the event in which the $i{:}th$ version of software is reliable to successfully achieve the given task at time t, (i=1,2,...,N). The software reliability function for the dependent N-version programming can then be written as

$$R(t) = \Pr\{\bigcup_{i=1}^{N} E_i(t)\} \qquad (9.43)$$

By using conditional probability, the events considered in the above equation can be decomposed into mutually exclusive events as

$$\begin{aligned} R(t) &= \Pr\{E_1(t)\} + \Pr\{E_2(t)\}\Pr\{\overline{E}_1(t)|E_2(t)\} + \\ &\cdots + \Pr\{E_N(t)\}\Pr\{\overline{E}_1(t), \overline{E}_2(t)\cdots\overline{E}_{N-1}(t)|E_N(t)\} \end{aligned} \qquad (9.44)$$

where $\Pr\{\overline{E}_1(t)|E_2(t)\}$ denotes the conditional probability that the first version of the software fails given that the second version of the software is reliable at time t.

Hence, each term in the software reliability expression of the above equation can be evaluated in terms of the probability of two distinct events. The first event indicates that the i:th version of software V_i is reliable while the second event indicates that all of its previous versions V_m (m=1,2,...,i-1) fails given that V_i is reliable.

The probability of the first event, $\Pr\{E_i(t)\}$, is straightforward. It can be calculated by multiplying the reliability functions of all the components that will make the i:th version V_i fail as

$$\Pr\{E_i(t)\} = \prod_{V_i \in M_{j,k}} R_{j,k}(t) \tag{9.45}$$

where $V_i \in M_{j,k}$ means that the i:th version of software V_i will fail if the component $M_{j,k}$ fails.

The probability of the second event, $\Pr\{\overline{E}_1(t), \overline{E}_2(t) \cdots \overline{E}_{i-1}(t) | E_i(t)\}$, is not as straightforward to compute. It can be done in the following steps:

Step 1: select all those components that can make any version(s) among the $V_1, V_2, \ldots, V_{i-1}$ fail while V_i is still reliable.

Step 2: use binary search tree (Johnsonbaugh, 2001) to find out all the exclusive combinations, which can make all the i-1 versions $V_1, V_2, \ldots, V_{i-1}$ fail among those components selected in step 1.

Step 3: add up all the probabilities of those exclusive combinations to obtain the probability of $\Pr\{\overline{E}_1(t), \overline{E}_2(t) \cdots \overline{E}_{i-1}(t) | E_i(t)\}$.

After computing $\Pr\{\overline{E}_1(t), \overline{E}_2(t) \cdots \overline{E}_{i-1}(t) | E_i(t)\}$ and $\Pr\{E_i(t)\}$, i=1,2,...,N, we can obtain the software reliability function for the dependent N-version programming by substituting them into Eq. (9.44). An example of aircraft landing is illustrated below.

Example 9.6. Suppose that three teams will compose three versions of a program to control the aircraft landing. If any one version is working, the aircraft can land successfully. These three versions may depend on each other through certain common cause failures. Those failures may occur on the common parts of some versions, such as using the same external electrical power, integrating the same software packages, sharing identical subroutines and so on.

As in the approaches presented above, the software is first decomposed into its individual components. As shown in example 9.5, The three dependent versions can be decomposed into 7 components corresponding to different common cause failures as shown in Fig. 9.9. $R_{j,k}$ denotes the reliability function of $M_{j,k}$. We have then

$$\begin{aligned} R(t) = {} & R_{11}R_{21}R_{23}R_{31} + R_{12}R_{21}R_{22}R_{31}(1 - R_{23}R_{11}) \\ & + R_{13}R_{22}R_{23}R_{31}[1 - R_{21}(R_{11} + R_{12} - R_{11}R_{12})] \end{aligned} \tag{9.46}$$

■

9.3.2. Optimal testing resource allocation

An optimization problem for testing resource allocation can be formulated to minimize the total cost for the N versions, when constrained by a fixed testing time budget T hours. Let t_i be the testing time allocated on the i:th version V_i (i=1,2,...,N), and the total testing time is less than T. The allocation of testing time significantly affects the total cost. There are mainly two parts in the cost:

(a) Test duration cost C_t: Here, the N versions of the software can be tested respectively given their allocated testing time t_i and their expected cost per unit of testing time c_i (i=1,2,...N). The test duration cost can be expressed as

$$C_t = \sum_{i=1}^{N} c_i t_i \tag{9.47}$$

where $c_i t_i$ is the expected cost in testing the i:th version.

(b) Risk cost C_r: this is the cost incurred by an unreliable system, see e.g. Pham and Zhang (1999). This can be expressed as

$$C_r = d(1 - R) \tag{9.48}$$

where d is the expected cost if the system fails and 1-R is the probability for the system to fail.

The total cost is the summation of the above two parts.

Denote by $t_{j,k}$ the testing time for component $M_{j,k}$. During the testing period, the component $M_{j,k}$ continues running and risks failure unless all the versions related to $M_{j,k}$ fail. Hence, the testing time of $t_{j,k}$ can be calculated by

$$t_{j,k} = \max_{m \in M_{jk}} (t_m) \qquad (9.49)$$

where $m \in M_{jk}$ means version m is related to component $M_{j,k}$. Hence, the reliability function of the component $M_{j,k}$ can be written as $R_{j,k}(x \mid t_{j,k})$ where x is the operation time after the test. The software reliability function $R(x \mid \vec{t})$ can then be derived through our approach presented above, where $\vec{t} = \{t_i | i = 1,2,...N\}$. The optimization problem to minimize the total cost by finding a set of testing time allocations $\vec{t}$, can be formulated by

$$\textbf{Minimize} \quad C(\vec{t}) = C_t + C_r = \sum_{i=1}^{N} c_i t_i + d[1 - R(x \mid \vec{t})] \qquad (9.50)$$

$$\textbf{Subject to:} \quad \sum_{i=1}^{N} t_i \leq T \qquad (9.51)$$

$$t_i \geq 0 \quad (i=1,2,\ldots,N) \qquad (9.52)$$

Solving this problem is also difficult, so heuristic algorithms need be implemented. An example is illustrated where a genetic algorithm is used here to solve it.

Example 9.7. Continuing with Example 9.6 (the air-craft landing example), suppose that the testing resource budget is 2000 hours of testing time, i.e., T=2000, that the testing cost per hour on the three versions are $c_1 = 0.3$, $c_2 = 0.2$, $c_3 = 0.28$, and that the risk cost $d = 10000$ if the aircraft cannot land successfully. The allocation problem becomes how to optimally allocate the 2000 hours on the three versions in order to minimize the total cost.

We assume that the common cause failures arriving on each component satisfy the Goel-Okumoto (GO) model. That is, the failure rate function for the components $M_{j,k}$ (j=1,2,3 and k=1,2,..., $K_{3,j}$) is modeled with:

$$\lambda_{j,k}(t) = a_{j,k} b_{j,k} \exp(-b_{j,k} t) \tag{9.53}$$

If the testing is stopped after t units of time, the reliability for a mission of duration t is given by (Yang & Xie, 2000)

$$R_{j,k}(x \mid t) = \exp\{-\lambda_{j,k}(t) \cdot x\} \tag{9.54}$$

The values of the parameters $a_{j,k}$ and $b_{j,k}$ in the GO model are given in Table 9.4 for this example.

Table 9.4. Parameters of the GO-model for each component.

Component	$M_{1,1}$	$M_{1,2}$	$M_{1,3}$	$M_{2,1}$	$M_{2,2}$	$M_{2,3}$	$M_{3,1}$
$a_{j,k}$	16.91	95.52	21.56	15.80	22.45	26.23	6.25
$b_{j,k}$	0.0059	0.0006	0.0041	0.0028	0.0021	0.0022	0.0056

Then, the reliability for the dependent three-version software can be obtained through Eq. (9.46). Substitute the parameters of Table 9.4 into Eq. (9.54) to compute the reliability functions of all the components, and then substitute them into Eq. (9.46) to compute the software reliability by assuming x=5 (i.e. it will take 5 hours for the aircraft to land).

To solve the optimization problem as Eqs. (9.50-9.52), a genetic algorithm is used to get the solution $\vec{t} = \{638.2, 1361.8, 0\}$. The best allocation of the 2000 hours should be to test: the first version for 638.2 hours; the second for 1361.8 hours and the third for 0 hour. The total expected cost $C(\vec{t})$=579.48 and the software reliability $R(5\,|\,\vec{t})$=0.988434.

■

9.4. Optimal Design of the Grid Architecture

9.4.1. Grid architecture design

For the grid computing systems (see Chapter 7), the network architecture is an important factor. Although the physical network may have already existed when building the grid, constructing a direct link between two remote nodes still lead to high cost where the direct link means that both nodes have the right to use the shared resources from each other. Hence, the cost of a direct link is mainly caused by preparing the resources, purchasing the right to use the resources, or dealing with the security problem during communication. Such cost is called link cost. Here a link can be a virtual link through the Internet/Intranet or even wireless.

On the other hand, if the grid computing system cannot complete the given tasks successfully (such as provide services), another kind of cost, called risk cost (Pham & Zhang, 1999) is caused by the unreliable computing. Hence, the total expected cost for designing the grid network architecture should consider both link cost and risk cost.

As Fig. 9.10, denoted by $\overrightarrow{PN_i}$ the set of programs executed by node i and $\overrightarrow{RN_i}$ the set of resources prepared in node i. Suppose the grid architecture (i.e.

the network of links among nodes) need be designed given $\vec{PN_i}$ and $\vec{RN_i}$ (i=1,2...N).

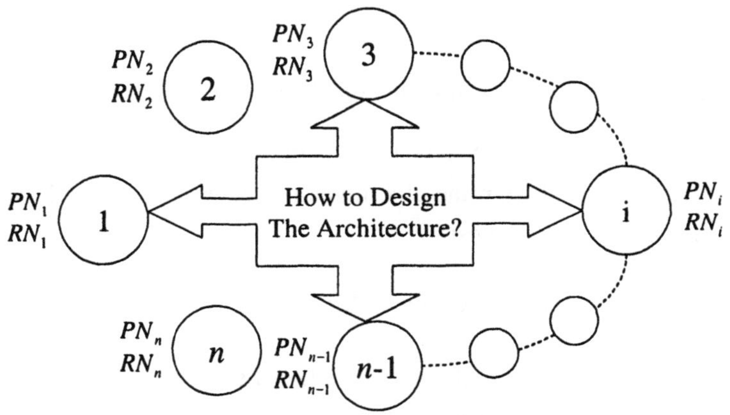

Fig. 9.10. Grid architecture design.

Adding more links among the nodes might increase the link cost but they could improve the system reliability to reduce the risk cost. In order to minimize the total cost, how to optimally design the network architecture of the grid, i.e. which link should exist or not, is important.

9.4.2. Optimization model

Denote by σ_{ij} the link between two nodes i and j ($i \neq j$). If σ_{ij}=1, there exists a link between the two nodes, and if σ_{ij}=0, there is no link between them. Here, σ is defined as a vector of $\{\sigma_{ij} \mid i, j \in [1, N], j > i\}$, which corresponds to a network architecture of N-node system. The length of σ is

$N(N$-1)/2, because the maximal number of links to completely connect all the N nodes is $N(N$-1)/2. Thus, the total link cost can be expressed as

$$D(\sigma)=\sum_{i=1}^{N}\sum_{j=i+1}^{N}\sigma_{ij}C_{ij} \tag{9.55}$$

in which C_{ij} is the cost to construct the link between node i and j.

The grid system reliability given a network architecture σ, denoted by $GSR(\sigma)$, can be derived from the algorithms presented in Chapter 7. We use the linear function of the risk cost as

$$C_r(\sigma)=C_R[1-GSR(\sigma)] \tag{9.56}$$

in which $1-GSR(\sigma)$ is the probability of a failed task of the grid and C_R is a constant which can be explained as the expected risk cost if the task fails.

The total expected cost is the summation of link cost and risk cost as

$$C(\sigma)=D(\sigma)+C_r(\sigma) \tag{9.57}$$

Our objective is to find an optimal network architecture σ^* to minimize the total expected cost. The optimization model is

$$\textbf{Minimize}\quad C(\sigma)=\sum_{i=1}^{N}\sum_{j=i+1}^{N}\sigma_{ij}C_{ij}+C_R[1-GSR(\sigma)] \tag{9.58}$$

$$\textbf{Subject to}\quad \sigma_{ij}=0 \text{ or } 1,\ \ i,j\in[1,N],\ j>i \tag{9.59}$$

If there are N nodes in the grid, the length of σ is $N(N$-1)/2 and the sample size of total network architectures is $2^{N(N-1)/2}$. Since the sample size increases exponentially to the number of nodes, it is difficult to exhaustively search all the samples for the optimal solution in a complex grid with many nodes.

Fortunately, for such complex grids, it is usually sufficient to find a good enough network design through certain heuristic algorithms, although it may not guarantee the optimum solution.

9.5. Optimal Integration of the Grid Services

9.5.1. Grid service integration problem

After the grid system is built, new services and resources are able to be further integrated into the grid by various virtual organizations. This is the objective of the second generation of the grid such as the Open Grid Service Architecture presented by Foster *et al.* (2001).

A grid service is to complete certain programs by using some resources distributed in the grid, as mentioned in Chapter 7. Hence, the integration of a new service on the grid is to allocate the programs and resources used by the service on certain reachable nodes of the grid. The reachable nodes represent those nodes that can be reached and used to upload the programs or integrate the resources of the new grid service.

In Chapter 7, we have analyzed the grid service reliability which is a special type of the grid system reliability by considering the programs of the given service in the grid. Recall that the grid service reliability is the probability that all the programs of a given service are achieved successfully under the grid computing environment. Maximizing it will serve as the objective of the service integration problem in this section.

The problem here is how to optimally allocate/integrate those programs and resources on the reachable nodes in order to maximize the service reliability after the integration. The organization may wonder how many program/resource redundancies they should prepare under the budget limitation, and how to distribute them on different reachable nodes of the grid system whose physical architecture has been constructed.

Suppose that a grid service is desired to complete M programs $P_1, P_2, ..., P_M$ which requires to access to H resources, $R_1, R_2, ..., R_h, ..., R_H$. These programs and resources are viewed as the components of the grid service. Denoted by C_i (i=1,2,...M,M+1,...,M+H) the i:th component of the service where the first M components corresponds to the M programs and the rest H components represent the H resources. The organization can prepare redundancies for each component (program or resource) but there is a budget constraint for the total cost denoted by B. Moreover, the number of redundancies for the i:th component should be no less than one and no more than an upper-bound denoted by U_i (i=1,2,...,K) where K=M+H (the total number of programs and resources in a grid service).

The organization can integrate the K components on N reachable nodes ($G_1, G_2, ..., G_N$) of the grid. Each node may have a limitation to integrate the components, such as the maximal number of components denoted by β_i (i=1,2,...,N).

Also, some components may have been fixed on some specific nodes, and some components are allowed to select some nodes to integrate or not allowed to integrate on other nodes. Hence, we use a_{ij} to describe the relationship between the i:th node (G_i) and the j:th component (C_j), i=1,2,...,N and j=1,2,...,K. Also, a_{ij} has three possible values (d, 0 and 1): if $a_{ij} = d$, it means the organization can freely choose whether to integrate the C_j on the G_i or not; if $a_{ij} = 0$, the C_j is not allowed to be integrated on the G_i; and if $a_{ij} = 1$, the C_j is fixed on the G_i. That is, α is defined as a vector of $\{ a_{ij} \mid i \in [1, N],\ j \in [1, K] \}$. The values of α can be determined given the relationship of the nodes and the service components.

Based on these conditions, the organization has to prepare the redundancies of each component and distribute them on those reachable nodes of the grid. To maximize the grid service reliability, the next section presents an optimization model for integrating a new service on the grid.

9.5.2. An optimization model

Let σ_{ij} represent the integration of the *j*:th component (C_j) on the *i*:th node (G_i): $\sigma_{ij}=0$ means that the C_j is not integrated with the G_i and $\sigma_{ij}=1$ means that C_j is integrated with G_i. Let σ be defined as a vector of $\{\sigma_{ij} \mid i\in[1,N],\ j\in[1,K]\}$ which represents an integration schedule of the grid service components on the nodes. Hence, given the structure of the grid, the grid service reliability can be only determined by the integration scheduling vector σ. The grid service reliability is then expressed by $R_S(\sigma)$, which can be computed by the algorithms in Chapter 7.3. Thus, the optimization problem becomes to find the optimal solution of σ so that the integrated grid service reliability is maximized. The optimization model is described as follows:

Decision variables: $\sigma=\{\sigma_{ij}=0,1 \mid i\in[1,N],\ j\in[1,K]\}$ (9.60)

Objective function: Maximize $R_S(\sigma)$ (9.61)

Constraints:

$$\sigma_{ij}=a_{ij}\ \text{(where } a_{ij}\neq d\text{)},\ i=1,2...N,\ j=1,2...K \tag{9.62}$$

$$1\le\sum_{i=1}^{N}\sigma_{ij}\le U_j \quad j=1,2...K \tag{9.63}$$

$$\sum_{j=1}^{K}\sigma_{ij}\le\beta_i,\ i=1,2...N \tag{9.64}$$

$$\sum_{i=1}^{N}\sum_{j=1}^{K}c_j\sigma_{ij}\le B \tag{9.65}$$

where the first constraint (9.62) is limited by a, the relationship of nodes and components:

if $a_{ij} = 1$ (i.e. the component C_j has to be fixed on G_i), the value of σ_{ij} has to be set to 1; and

if $a_{ij} = 0$ (i.e. the component C_j cannot be integrated on G_i), the value of σ_{ij} has to be set to 0.

In the second constraint, Eq. (9.63), $\sum_{i=1}^{N} \sigma_{ij}$ represents the total number of redundancies of component C_j integrated in the grid, which should between 1 and its upper-bound U_j. In the third constraint, Eq. (9.64), $\sum_{j=1}^{K} \sigma_{ij}$ represents the total number of components integrated on the node G_i, which ought to be no more than its upper-bound β_i. Finally, for Eq. (9.65), c_j is the cost to prepare a redundancy of the *j*:th component, so that the left hand side represents the total cost for the integration of the grid service, which has to be no more than the budget *B*.

In order to solve this optimization problem, heuristic algorithms can be used.

9.6. Notes and References

For the optimal number of redundant units, many other studies have also been presented. Pham (1992) determined the optimal number of spare units that minimize the average total system cost. Imaizumi *et al.* (2000) obtained the mean time and the expected cost until system failure and discussed an optimal number which minimizes the expected cost for a system with multiple microprocessor units. Hsieh & Hsieh (2003) developed a relationship between system cost and hardware redundancy levels, and presented an optimization model aiming at minimizing the total system cost. Hsieh (2003) further presented optimization models for the policies of task allocation and hardware redundancy of the distributed computing systems. Chang *et al.* (2003) further presented an optimization model in dynamically adding and removing redundant units of the computing system.

Optimal testing-resource allocation problem has also been studied extensively in the literature. Yamada & Nishiwaki (1995) proposed optimal allocation policies for testing-resource based on a software reliability growth models. Based on the hyper-geometric distribution software reliability growth model, Hou *et al.* (1996) investigated two optimal resource allocation problems in software module testing. Leung (1997) later studied the dynamic resource-allocation for software-module testing. Coit (1998) presented a method to allocate subsystem reliability growth test time in order to maximize the system reliability when designers are confronted with limited testing resources. Lyu *et al.* (2002) further considered software component testing resource allocation for a system with single or multiple applications. For the networked system, Hsieh & Lin (2003) aimed to determine the optimal resource allocation policy at source nodes subject to given resource demands at sink nodes such that the network reliability of the stochastic-flow network is maximized.

For the grid computing system design, Buyya *et al.* (2002) presented two optimization strategies in providing grid services based on the economic models of the resource management. Furmento *et al.* (2002) used composite performance models to optimally combine currently available component into the network of the Grid environment. According to the QoS requirements, Dogan & Ozguner (2002) presented the optimization model for scheduling independent tasks in grid computing with time-varying resource prices.

Optimization models have been widely studied in other areas of the computing systems. Okumoto & Goel (1980) first discussed the software optimal release policy from the cost-benefit viewpoint. There are many follow-up papers and a chapter in Xie (1991) is devoted to this issue. Zheng (2002) considered some dynamic release policies. The sensitivity of software release time remains an issue and some preliminary discussion can be found in Xie & Hong (1998). Quigley & Walls (2003) discussed the confidence intervals for reliability-growth models when the sample-size is small which is a common situation.

Ashrafi *et al.* (1994) discussed the optimal design of *N*-version programming system. Berman & Kumar (1999) considered some optimization models for recovery block design. Jung & Choi (1999) studied some optimization models for modular systems based on cost analysis.

Tom & Murthy (1999) implemented graph matching and state space search techniques in optimizing the schedule of task allocation on the distributed computing systems. Karatza (2001) investigated optimal scheduling policies in a heterogeneous distributed system, where half of the total processors have double the speed of the others. Kuo & Prasad (2000) reviewed some system-reliability optimization models. Kuo & Zuo (2003) recently summarized many reliability optimization models in the computing systems.

References

Akhtar, S. (1994), Reliability of k-out-of-n:G systems with imperfect fault-coverage, *IEEE Transactions on Reliability*, **43**, 101-106.

Aki, S. and Hirano, K. (1996), Lifetime distribution and estimation problems of consecutive-k-out-of-n:F systems, *Annals of the Institute of Statistical Mathematics*, **48** (1), 185-199.

Alexopoulos, C. and Shultes, B.C. (2001), Estimating reliability measures for highly-dependable Markov systems using balanced likelihood ratios, *IEEE Transactions on Reliability*, **50**, 265-280.

Ammann, P.E. and Knight, J.C. (1988), Data diversity: an approach to software fault tolerance, *IEEE Transactions on Computers*, **37** (4), 418–425.

Ammar, H.H., Cukic, B., Mili, A. and Fuhrman, C. (2000), A comparative analysis of hardware and software fault tolerance: impact on software reliability engineering, *Annals of Software Engineering*, **10**, 103-150.

Arulmozhi, G. (2003), Direct method for reliability computation of k-out-of-n:G systems, *Applied Mathematics and Computation*, **143** (2-3), 421-429.

Ashrafi, N., Berman, O. and Cutler, M. (1994), Optimal design of large software-systems using N-version programming, *IEEE Transactions on Reliability*, **43**, 344-350.

Avizienis, A. (1985), The N-version approach to fault tolerant software, *IEEE Transactions on Software Engineering*, **11**, 1491-1501.

Barlow, R.E. and Proschan, F. (1981), *Statistical Theory of Reliability and Life Testing: Probability Models*, Silver Spring, MD: To Begin With.

Becker, G., Camarinopoulos, L. and Zioutas, G. (2000), A semi-Markovian model allowing for inhomogenities with respect to process time, *Reliability Engineering & System Safety*, **70** (1), 41-48.

Belli, F. and Jedrzejowicz, P. (1991), An approach to the reliability optimization of software with redundancy, *IEEE Transactions on Software Engineering*, **17** (3), 310-312.

Berman, O. and Kumar, U.D. (1999), Optimization models for recovery block schemes, *European Journal of Operational Research*, **115**, 368-379.

Blischke, W.R. and Murthy, D.N.P. (2000), *Reliability*, New York: Wiley.

Bobbio, A., Premoli, A. and Saracco, O. (1980), Multi-state homogeneous Markov models in reliability analysis, *Microelectronics and Reliability*, **20** (6), 875-880.

Boland, P.J. and Singh, H. (2003), A birth-process approach to Moranda's geometric software-reliability model, *IEEE Transactions on Reliability*, **52**, 168-174.

Brunelle, R.D. and Kapur, K.C. (1999), Review and classification of reliability measures for multistate and continuum models, *IIE Transactions*, **31** (12), 1171-1180.

Bukowski, J.V. and Goble, W.M. (2001), Defining mean time-to-failure in a particular failure-state for multi-failure-state systems, *IEEE Transactions on Reliability*, **50**, 221-228.

Buyya, R., Abramson, D., Giddy, J. and Stockinger, H. (2002), Economic models for resource management and scheduling in Grid computing, *Concurrency and Computation Practice & Experience*, **14** (13-15), 1507-1542.

Buyya, R., Branson, K., Giddy, J. and Abramson, D. (2003), The virtual laboratory: a toolset to enable distributed molecular modelling for drug design on the world-wide grid, *Concurrency and Computation Practice & Experience*, **15** (1), 1-25.

Cao, J., Jarvis, S.A., Saini, S., Kerbyson, D.J. and Nudd, G.R. (2002), ARMS: An agent-based resource management system for grid computing, *Scientific Programming*, **10** (2), 135-148.

Casavant, T.L. and Singhal, M. (1994), *Readings in Distributed Computing Systems*, Los Alamitos, CA: IEEE Computer Society Press.

Chang, C.W.J., Hsiao, M.F. and Marek-Sadowska, M. (2003), A new reasoning scheme for efficient redundancy addition and removal, *IEEE Transactions on Computer-Aided Design of Integrated Circuits and Systems*, **22** (7), 945-951.

Chang, G.J., Cui, L.R. and Hwang, F.K. (2000), *Reliabilities of Consecutive-k Systems*, Dordrecht, Netherlands: Kluwer Academic Publishers.

Chang, M.S., Chen, D.J., Lin, M.S. and Ku, K.L. (2000), The distributed program reliability analysis on star topologies, *Computers & Operations Research*, **27**, 129-142.

Chen, D.J., Chang, M.S., Sheng, M.C. and Horng, M.S. (1998), Time-constrained distributed program reliability analysis, *Journal of Information Science and Engineering*, **14** (4), 891-911.

Chen, D.J., Chen, R.S. and Huang, T.H. (1997), A heuristic approach to generating file spanning trees for reliability analysis of distributed computing systems, *Computers and Mathematics with Application*, **34**, 115-131.

Chen, D.J. and Huang, T.H. (1992), Reliability analysis of distributed systems based on a fast reliability algorithm, *IEEE Transactions on Parallel and Distributed Systems*, **3** (2), 139-154.

Chen, I.R. and Bastani, F.B. (1992), Reliability of fully and partially replicated systems, *IEEE Transactions on Reliability*, **41** (2), 175-182

Chen, I.R. and Bastani, F.B. (1994), Warm standby in hierarchically structured process-control programs, *IEEE Transactions on Software Engineering*, **20** (8), 658-663.

Cheung, R.C. (1980), A user-oriented software reliability model, *IEEE Transactions on Software Engineering*, **6** (2), 118-125.

Chiu, C.C., Yeh, Y.S. and Chou, J.S. (2002), A fast algorithm for reliability-oriented task assignment in a distributed system, *Computer Communications*, **25** (17), 1622-1630.

Choi, J.G. and Seong, P.H. (2001), Dependability estimation of a digital system with consideration of software masking effects on hardware faults, *Reliability Engineering & Systems Safety*, **71** (1), 45-55.

Clemen, R.T. (1995), *Making Hard Decisions: An Introduction to Decision Analysis*, New Jersey: Duxbury Press.

Coit, D.W. (1998), Economic allocation of test times for subsystem-level reliability growth testing. *IIE Transactions*, **30** (12), 1143-1151.

Coit, D.W. and Smith, A.E. (1996), Reliability optimization of series-parallel systems using a genetic algorithm, *IEEE Transactions on Reliability*, **45** (2), 254-266.

Dai, Y.S., Xie, M. and Poh, K.L. (2002), Reliability analysis of Grid computing systems, *Proceedings of the Pacific Rim International Symposium on Dependable Computing*, IEEE Computer Society Press, pp. 97-103.

Dai, Y.S., Xie, M., Poh, K.L. and Liu, G.Q. (2003a), A study of service reliability and availability for distributed systems, *Reliability Engineering & System Safety*, **79** (1), 103-112.

Dai, Y.S., Xie, M., Poh, K.L. and Yang, B. (2003b), Optimal testing-resource allocation with genetic algorithm for modular software systems, *Journal of Systems and Software*, **66** (1), 47-55.

Das, O. and Woodside, C.M. (2001), Evaluating layered distributed software systems with fault-tolerant features, *Performance Evaluation*, **45**, 57-76.

Dogan, A. and Ozguner, F. (2002), Scheduling independent tasks with QoS requirements in grid computing with time-varying resource prices, *Lecture Notes of Computer Science*, **2536**, 58-69.

Duane, J.T. (1964), Learning curve approach to reliability monitoring, *IEEE Transactions on Aerospace*, **2**, 563-566.

Dugan, J.B. and Lyu, M.R. (1994), System reliability analysis of an N-version programming application, *IEEE Transactions on Reliability*, **43** (4), 513-519.

Dugan, J.B. and Lyu, M.R. (1995), System-level reliability and sensitivity analyses for three fault-tolerant system architectures, *Dependable Computing for Critical Applications*, **4**, 459-477.

Elsayed, E.A. (1996), *Reliability Engineering*, Reading, MA: Addison Wesley.

Fahmy, H.M.A. (2001), Reliability evaluation in distributed computing environments using the AHP, *Computer Networks*, **36** (5-6): 597-615.

Fakhre-Zakeri, I. and Slud, E. (1995), Mixture models for reliability of software with imperfect debugging: identifiability of parameters, *IEEE Transactions on Reliability*, **44**, 104-113.

Fleming, K. and Silady, F. (2002), A risk informed defense-in-depth framework for existing and advanced reactors, *Reliability Engineering & System Safety*, **78**, 205-225.

Foster, I. and Kesselman, C. (1998), *The Grid: Blueprint for a New Computing Infrastructure*, San Francisco, CA: Morgan-Kaufmann.

Foster, I., Kesselman, C. and Tuecke, S. (2001), The anatomy of the Grid: Enabling scalable virtual organizations, *International Journal of High Performance Computing Applications*, **15** (3), 200-222.

Foster, I., Kesselman, C., Nick, J.M. and Tuecke, S. (2002), Grid services for distributed system integration, *Computer*, **35** (6), 37-46.

Frey, J., Tannenbaum, T., Livny, M., Foster, I. and Tuecke, S. (2002), Condor-G: a computation management agent for multi-institutional grids, *Cluster Computing*, **5** (3), 237-246.

Fricks, R.M., Puliafito, A. and Trivedi, K.S. (1999), Performance analysis of distributed real-time databases, *Performance Evaluation*, **35**, 145-169.

Fryer, M.O. (1985), Risk assessment of computer controlled systems, *IEEE Transactions on Software Engineering*, **11** (1), 125-129.

Furmento, N., Mayer, A., McGough, S., Newhouse, S., Field, T. and Darlington, J. (2002), ICENI: Optimisation of component applications within a Grid environment, *Parallel Computing*, **28** (12), 1753-1772.

Gaudoin, O., Lavergne, C. and Soler, J.L. (1994), A generalized geometric de-eutrophication software-reliability model, *IEEE Transactions on Reliability*, **43** (4), 536-541.

Gaudoin, O. and Soler, J.L. (1992), Statistical analysis of the geometric de-eutrophication software-reliability model, *IEEE Transactions on Reliability*, **41** (4), 518-524.

Gnedenko, B. and Ushakov, I. (1995), *Probabilistic Reliability Engineering*, New York: Wiley.

Goel, A.L. (1980), A summary of the discussion on "an analysis of competing software reliability models", *IEEE Transactions on Software Engineering*, **6**, 501-502.

Goel, A.L. (1985), Software reliability models: assumptions, limitations, and applicability, *IEEE Transactions on Software Engineering*, **11**, 1411-1423.

Goel, A.L. and Okumoto, K. (1979), Time dependent error-detection rate model for software reliability and other performance measures, *IEEE Transactions on Reliability*, **28**, 206-211.

Goel, A.L. and Soenjoto, J. (1981), Models for hardware-software system operational performance evaluation, *IEEE Transactions on Reliability*, **30**, 232-239.

Goseva-Popstojanova, K. and Trivedi, K.S. (2000), Failure correlation in software reliability model, *IEEE Transactions on Reliability*, **49**, 37-48.

Goseva-Popstojanova, K. and Trivedi, K.S. (2003), Architecture-based approaches to software reliability prediction, *Computers and Mathematics with Applications*, **46** (7), 1023-1036.

Hariri, S. and Mutlu, H. (1995), Hierarchical modeling of availability in distributed systems, *IEEE Transactions on Software Engineering*, **21**, 50–56.

Hecht, H. and Hecht, M. (1986), Software reliability in the system context, *IEEE Transactions on Software Engineering*, **12** (1), 51-58.

Helander, M.E., Zhao, M. and Ohlsson, N. (1998), Planning models for software reliability and cost, *IEEE Transactions on Software Engineering*, **24** (6), 420-434.

Hillier, F.S. and Lieberman, G.J. (1995), *Introduction to Operations Research*, New York: McGraw-Hill.

Hou, R.H., Kuo, S.Y. and Chang, Y.P. (1996), Needed resources for software module test using the hyper-geometric software reliability growth model, *IEEE Transactions on Reliability*, **45** (4), 541-549.

Hoyland, A. and Rausand, M. (1994), *System Reliability Theory*, New York: Wiley.

Hsieh, C.C. (2003), Optimal task allocation and hardware redundancy policies in distributed computing systems, *European Journal of Operational Research*, **147** (2), 430-447.

Hsieh, C.C. and Hsieh, Y.C. (2003) Reliability and cost optimization in distributed computing systems, *Computers & Operations Research*, **30** (8), 1103-1119.

Hsieh, C.C. and Lin, M.H. (2003), Reliability-oriented multi-resource allocation in a stochastic-flow network, *Reliability Engineering & System Safety*, **81** (2), 155-161.

Huang, C.Y. and Kuo, S.Y. (2003), Analysis of incorporating logistic testing-effort function into software reliability modeling, *IEEE Transactions on Reliability*, **51** (3), 261-270.

Huang, C.Y., Lyu, M.R. and Kuo, S.Y. (2003), A unified scheme of some nonhomogenous Poisson process models for software reliability estimation, *IEEE Transactions on Software Engineering*, **29** (3), 261-269.

Hussain, D.S. and Hussain. K.M. (1992), *Information Management: Organization, Management, and Control of Computer Processing*, New York: Prentice Hall.

Imaizumi, M., Yasui, K. and Nakagawa, T. (2000), An optimal number of microprocessor units with watchdog processor, *Mathematical and Computer Modelling*, **31** (10-12), 183-189.

Jelinski, Z. and Moranda, P.B. (1972), Software reliability research, In: Freiberger W. (ed), *Statistical Computer Performance Evaluation*, New York: Academic Press, pp. 465-497.

Johnsonbaugh, R. (2001), *Discrete Mathematics*, New Jersey: Prentice-Hall.

Jung, H.W. and Choi, B.J, (1999), Optimization models for quality and cost of modular software, *European Journal of Operational Research*, **112**, 613-619.

Kanoun, K. and Ortalo-Borrel, M. (2000), Fault-tolerant system dependability - Explicit modeling of hardware and software component-interactions, *IEEE Transactions on Reliability*, **49** (4), 363-376.

Kapur, P.K., Garg, R.B. and Kumar, S. (1998), *Contributions to Hardware and Software Reliability*, Singapore: World Scientific.

Karatza, H.D. (2001), Job scheduling in heterogeneous distributed systems, *Journal of Systems and Software*, **56** (3), 203-212.

Kaufman, G.M. (1996), Successive sampling and software reliability, *Journal of Statistical Planning and Inference*, **49** (3), 343-369.

Ke, W.J. and Wang, S.D. (1997), Reliability evaluation for distributed computing networks with imperfect nodes, *IEEE Transactions on Reliability*, **46** (3), 342-349.

Keahey, K., Fredian, T., Peng, Q., Schissel, D.P., Thompson, M., Foster, I., Greenwald, M. and McCune, D. (2002), Computational grids in action: the national fusion collaboratory, *Future Generation Computer Systems*, **18** (8), 1005-1015.

Keene, S. and Lane, C. (1992), Combined hardware and software aspects of reliability, *Quality and Reliability Engineering International*, **8** (5), 419-426.

Kijima, M. (1997), *Markov Processes for Stochastic Modeling*, New York: Chapman & Hall.

Kim, K.H. and Welch, H.O. (1989), Distributed execution of recovery blocks: an approach for uniform treatment of hardware and software faults in real-time applications, *IEEE Transactions on Computers*, **38**, 626-636.

Knight, J.C. and Leveson, N.G. (1986), An experimental evaluation of the assumption of independence in multiversion programming, *IEEE Transactions on Software Engineering*, **12**, 96-109.

Kolowrocki, K. (2001), On limit reliability functions of large multi-state systems with ageing components, *Applied Mathematics and Computation*, **121** (2-3), 313-361.

Krauter, K., Buyya, R. and Maheswaran, M. (2002), A taxonomy and survey of grid resource management systems for distributed computing, *Software - Practice and Experience*, **32** (2), 135-164.

Kremer, W. (1983), Birth-death and bug counting (software reliability), *IEEE Transactions on Reliability*, **32** (1), 37-47.

Kubat, P. (1989), Assessing reliability of modular software, *Operation Research Letters*, **8**, 35-41.

Kumar, A. and Malik, K. (1991), Voting mechanisms in distributed systems, *IEEE Transactions on Reliability*, **40** (5), 593-600.

Kumar, A. and Agrawal, D.P. (1993), A generalized algorithm for evaluating distributed-program reliability, *IEEE Transactions on Reliability*, **42**, 416-424.

Kumar, A. and Agrawal, D.P. (1996), Parameters for system effectiveness evaluation of distributed systems, *IEEE Transactions on Computers*, **45** (6), 746-752.

Kumar, A., Rai, S. and Agarwal, D.P. (1988), On computer communication network reliability under program execution constraints, *IEEE Transactions on Selected Areas in Communications*, **6**, 1393-1400.

Kumar, V.K., Hariri, S. and Raghavendra, C.S. (1986), Distributed program reliability analysis, *IEEE Transactions on Software Engineering*, **12**, 42-50.

Kuo, S.Y., Huang, C.Y. and Lyu, M.R. (2001), Framework for modeling software reliability, using various testing-efforts and fault-detection rates, *IEEE Transactions on Reliability*, **50**, 310-320.

Kuo, W. and Prasad, V.R. (2000), An annotated overview of system-reliability optimization, *IEEE Transactions on Reliability*, **49**, 176-187.

Kuo, W. and Zuo, M.J. (2003), *Optimal Reliability Modeling: Principles and Applications*, New York: Wiley.

Lai, C.D., Xie, M., Poh, K.L., Dai, Y.S. and Yang, P. (2002), A model for availability analysis of distributed software/hardware systems, *Information and Software Technology*, **44** (6), 343-350.

Langer, A.M. (2000), *Analysis and Design of Information Systems*, New York: Springer.

Lanus, M., Yin, L. and Trivedi, K.S. (2003), Hierarchical composition and aggregation of state-based availability and performability models, *IEEE Transactions on Reliability*, **52** (1), 44-52.

Laprie, J.C. and Kanoun, K. (1992), X-ware reliability and availability modeling, *IEEE Transactions on Software Engineering*, **18**, 130-147.

Laprie, J.C., Arlat, J., Biounes, C. and Kanoun, K. (1990), Definition and analysis of hardware and software-fault-tolerant architectures, *Computer*, **23** (7), 39-51.

Latif-Shabgahi, G., Bennett, S. and Bass, J.M. (2000), Empirical, evaluation of voting algorithms used in fault-tolerant control systems, *Parallel and Distributed Computing and Systems*, **1**, 340-345.

Ledoux, J. (1999), Availability modeling of modular software, *IEEE Transactions on Reliability,* **48** (2), 159–168.

Leung, Y.W. (1997), Dynamic resource-allocation for software-module testing, *Journal of Systems and Software,* **37** (2), 129-139.

Levitin, G. (2001), Analysis and optimization of weighted voting systems consisting of voting units with limited availability, *Reliability Engineering & System Safety*, **73** (1), 91-100.

Levitin, G. (2002), Asymmetric weighted voting systems, *Reliability Engineering & System Safety*, **76** (2), 205-212.

Levitin, G. (2003), Optimal multilevel protection in series-parallel systems, *Reliability Engineering & System Safety*, **81** (1), 93-102.

Levitin, G., Dai, Y.S., Xie M. and Poh, K.L (2003), Optimizing survivability of multi-state systems with multi-level protection by multi-processor genetic algorithm, *Reliability Engineering & System Safety*, **82**, 93-104.

Levitin, G. and Lisnianski, A. (2001), A new approach to solving problems of multi-state system reliability optimization, *Quality and Reliability Engineering International*, **17** (2), 93-104.

Levitin, G. and Lisnianski, A. (2003), Optimizing survivability of vulnerable series-parallel multi-state systems, *Reliability Engineering & System Safety*, **79** (3), 319-331.

Levitin, G., Lisnianski, A., Beh-Haim, H. and Elmakis, D. (1998), Redundancy optimization for series-parallel multi-state systems, *IEEE Transactions on Reliability*, **47**, 165-172.

Limnios, N. (1997), Dependability analysis of semi-Markov systems, *Reliability Engineering & System Safety,* **55** (3), 203-207.

Limnios, N. and Oprisan, G. (2000), *Semi-Markov Processes and Reliability*, Boston: Birkhauser.

Lin, M.S. (2001), Linear-time algorithms for computing the reliability of bipartite and (# <= 2) star distributed computing systems, *Computers & Operations Research*, **30** (11), 1697-1712.

Lin, M.S., Chang, M.S. and Chen, D.J. (1999a), Distributed-program reliability analysis: complexity and efficient algorithms, *IEEE Transactions on Reliability*, **48**, 87-95.

Lin, M.S., Chang, M.S. and Chen, D.J. (1999b), Efficient algorithms for reliability analysis of distributed computing systems, *Information Sciences*, **117** (1-2), 89-106.

Lin, M.S., Chang, M.S., Chen, D.J. and Ku, K.L. (2001), The distributed program reliability analysis on ring-type topologies, *Computers & Operations Research*, **28**, 625-635.

Lin, M.S. and Chen, D.J. (1997), The computational complexity of the reliability problem on distributed systems, *Information Processing Letters*, **64**, 143-147.

Lin, M.S., Chen, D.J. and Hong, M.S. (1999), The reliability analysis of distributed computing systems with imperfect nodes, *The Computer Journal*, **42** (2), 129-141.

Lisnianski, A. and Levitin, G. (2003), *Multi-state System Reliability*, Singapore: World Scientific.

Littlewood, B. (1975), A reliability model for systems with Markov structure, *Applied Statistics*, **24** (2), 172–177.

Littlewood, B. (1979), How to measure software reliability and how not to, *IEEE Transactions on Reliability*, **28** (2), 103-110.

Littlewood, B. (1984), Rationale for a modified Duane model, *IEEE Transactions on Reliability*, **33** (2), 157-159.

Littlewood, B., Popov, P. and Strigini, L. (2002), Assessing the reliability of diverse fault-tolerant software-based systems, *Safety Science*, **40** (9), 781-796.

Littlewood, B. and Verrall, J.L. (1981), Likelihood function of a debugging model for computer software reliability, *IEEE Transactions on Reliability*, **30**, 145-148.

Liu, P.X., Zuo, M.J. and Meng, M.Q.H. (2003), Using neural network function approximation for optimal design of continuous-state parallel–series systems, *Computers & Operations Research*, **30** (3), 339-352.

Livny, M. and Raman, R. (1998), High-throughput resource management, In *The Grid: Blueprint for a New Computing Infrastructure*, San Francisco, CA: Morgan-Kaufmann, pp. 311-338.

Lopez-Benitez, N. (1994), Dependability modeling and analysis of distributed programs, *IEEE Transactions on Software Engineering*, **20** (5), 345-352.

Loy, D., Dietrich, D. and Schweinzer, H.J. (2001), *Open Control Networks*, Boston, MA: Kluwer Academic Publishers.

Lyu, M.R. (1996), *Handbook of Software Reliability Engineering*, IEEE Computer Society Press, New York: McGraw-Hill.

Lyu, M.R., Rangarajan, S. and van Moorsel, A.P.A. (2002), Optimal allocation of test resources for software reliability growth modeling in software development, *IEEE Transactions on Reliability*, **51** (2), 183-192.

Mahmood, A. (2001), Task allocation algorithms for maximizing reliability of heterogeneous distributed computing systems, *Control and Cybernetics*, 30 (1), 115-130.

Malluhi, Q.M. and Johnston, W.B. (1998), Coding for high availability of a distributed-parallel storage system, *IEEE Transactions on Parallel and Distributed Systems*, **9**, 1237-1252.

Mendiratta, V.B. (1998), Reliability analysis of clustered computing systems, *Proceedings of the 9th International Symposium on Software Reliability Engineering*, pp. 268-272.

Miller, D.R. (1986), Exponential order statistic models of software reliability growth, *IEEE Transactions on Software Engineering*, **12** (1), 12-24.

Moranda, P.B. (1979), Event-altered rate models for general reliability analysis, *IEEE Transactions on Reliability*, **28** (5), 376-381.

Musa, J.D. (1998), *Software Reliability Engineering: More Reliable Software, Faster Development and Testing*, New York: McGraw-Hill.

Musa, J.D., Iannino, A. and Okumoto, K. (1987), *Software Reliability: Measurement, Prediction, Application*, New York: McGraw-Hill.

Musa, J.D. and Okumoto, K. (1984), A logarithmic Poisson execution time model for software reliability measurement, *Proceedings of the 7th International Conference on Software Engineering*, pp. 230-238.

Okumoto, K. and Goel, A.L. (1980), Optimum release time for software systems based on reliability and cost criteria, *Journal of Systems and Software*, **1**, 315-318.

Ozekici, S. and Soyer, R. (2003), Reliability of software with an operational profile, *European Journal of Operational Research*, **149** (2), 459-474.

Pasquini, A., Pistolesi, G. and Rizzo, A. (2001), Reliability analysis of systems based on software and human resources, *IEEE Transactions on Reliability*, **50** (4), 337-345.

Pham, H. (1992), Optimal design of *k*-out-of-*n* redundant systems, *Microelectronics and Reliability*, **32** (1-2), 119-126.

Pham, H. (1997), Reliability analysis of digital communication systems with imperfect voters, *Mathematical and Computer Modelling*, **26**, 103-112.

Pham, H. (2000), *Software Reliability*, Singapore: Springer-Verlag.

Pham, H. (2003), Software reliability and cost models: Perspectives, comparison, and practice, *European Journal of Operational Research*, **149** (3), 475-489.

Pham, H., Nordmann, L. and Zhang, X.M. (1999), General imperfect-software-debugging model with S-shaped fault-detection rate, *IEEE Transactions on Reliability*, **48**, 169-175.

Pham, H., Suprasad, A. and Misra, R.B. (1997), Availability and mean life time prediction of multistage degraded system with partial repairs, *Reliability Engineering & System Safety*, **56** (2), 169-173.

Pham, H. and Zhang, X.M. (1999), A software cost model with warranty and risk cost, *IEEE Transactions on Computers*, **48** (1), 71-75.

Pham, H. and Zhang, X.M. (2003), NHPP software reliability and cost models with testing coverage, *European Journal of Operational Research*, **145** (2), 443-454.

Pierre, S. and Hoang, H.H. (1990), An artificial intelligence approach for improving computer communications network topologies, *Journal of Operational Research Society*, **41** (5), 405-418.

Postel, J. and Touch, J. (1998), Network infrastructure, In *The Grid: Blueprint for a New Computing Infrastructure*, San Francisco: Morgan-Kaufmann, pp. 533-566.

Pourret, O., Collet, J. and Bon, J.L. (1999), Evaluation of the unavailability of a multistate-component system using a binary model, *Reliability Engineering & System Safety*, **64** (1), 13-17.

Prasad, V.B. (1991), Markovian model for the evaluation of reliability of computer networks with intermittent faults, *Proceedings of the 1991 IEEE International Symposium on Circuits and Systems*, pp. 2084-2087.

Pukite, P. and Pukite, J. (1998), *Modeling for Reliability Analysis*. New York: IEEE Press.

Quigley, J. and Walls, L. (2003), Confidence intervals for reliability-growth models with small sample-sizes, *IEEE Transactions on Reliability*, **52** (2), 257-262.

Rajgopal, J. and Mazumdar, M. (2002), Modular operational test plans for inferences on software reliability based on a Markov model, *IEEE Transactions on Software Engineering*, **28** (4), 358-363.

Ross, S.M. (2000), *Introduction to Probability Models*, San Diego, CA: Academic Press.

Sahinoglu, M., Deely, J.J. and Capar, S. (2001), Stochastic Bayes measures to compare forecast accuracy of software-reliability models, *IEEE Transactions on Reliability*, **50** (1), 92-97.

Schick, G.J. and Wolverton, R.W. (1978), An analysis of competing software reliability models, *IEEE Transactions on Software Engineering*, **4**, 104-120.

Schneidewind, N.F. (1975), Analysis of error processes in computer software, *Sigplan Notices*, **10**, 337-346.

Shanthikumar, J.G. (1981), Ageneral software reliability model for performance prediction, *Microelectronics and Reliability*, **23**, 903-943.

Shao, J. and Lamberson, L.R. (1991), Modeling a shared-load *k*-out-of-*n*:G system, *IEEE Transactions on Reliability*, **40** (2), 205-209.

Shooman, M.L. (1990), *Probabilistic Reliability: An Engineering Approach*. Florida: Robert E. Krieger Publishing.

Shyur, H.J. (2003), A stochastic software reliability model with imperfect-debugging and change-point, *Journal of Systems and Software*, **66** (2), 135-141.

Siegrist, K. (1988), Reliability of systems with Markov transfer of control, *IEEE Transactions on Software Engineering*, **14** (10), 1478-1480.

Sridharan, V. and Jayashree, P.R. (1998), Transient solutions of a software model with imperfect debugging and generation of errors by two servers, *Mathematical and Computer Modelling*, **27**, 103-108.

Sumita, U. and Masuda, Y. (1986), Analysis of software availability/reliability under the influence of hardware failures, *IEEE Transactions on Software Engineering*, **12**, 32-41.

Tokuno, K. and Yamada, S. (2000), An imperfect debugging model with two types of hazard rates for software reliability measurement and assessment, *Mathematical and Computer Modelling*, **31** (10-12), 343-352

Tokuno, K. and Yamada, S. (2001), Markovian modeling for software availability analysis under intermittent use, *International Journal of Reliability, Quality and Safety Engineering*, **8** (3), 249-258.

Tom, P.A. and Murthy, C.S.R. (1999), Optimal task allocation in distributed systems by graph matching and state space search, *Journal of Systems and Software*, **46**, 59-75.

Trivedi, K.S. (1982), *Probability and Statistics with Reliability, Queuing, and Computer Applications*, Englewood, NJ: Prentice-Hall.

Valiant, L.G. (1979), The complexity of enumeration and reliability problems, *SIAM Journal of Computing*, **8**, 410-421.

Weissman, J.B. and Lee, B.D. (2002), The virtual service grid: An architecture for delivering high-end network services, *Concurrency Computation Practice and Experience*, **14** (4), 287-319.

Welke, S.R., Johnson, B.W. and Aylor, J.H. (1995), Reliability modeling of hardware/software systems, *IEEE Transactions on Reliability*, **44** (3), 413-418.

Wu, S.M. and Chan, L.Y. (2003), Performance utility-analysis of multi-state systems, *IEEE Transactions on Reliability*, **52** (1), 14-21.

Xie, M. (1987), A shock model for software failures, *Microelectronics and Reliability*, **27**, 717-724.

Xie, M. (1991), *Software Reliability Modelling*, Singapore: World Scientific.

Xie, M. (2000), Software reliability models - past, present and future. In *Recent Advances in Reliability Theory: Methodology, Practice, and Inference*, Eds. N. Limnios and M. Nikulin, Boston: Birkhäuser, pp. 325-340.

Xie, M. and Hong, G.Y. (1998), A study of the sensitivity of software release time, *Journal of Systems and Software*, **44** (2), 163-168.

Xie, M., Hong, G.Y. and Wohlin, C. (1997), A study of the exponential smoothing technique in software reliability growth prediction, *Quality and Reliability Engineering International*, **13** (6), 347-353.

Xie, M. and Yang, B. (2003), A study of the effect of imperfect debugging on software development cost, *IEEE Transactions on Software Engineering*, **29** (5), 471-473.

Xie, M. and Zhao, M. (1993), On some reliability growth models with graphical interpretations, *Microelectronics and Reliability*, **33** (2), 149-167.

Xue, J. and Yang, K. (1995), Dynamic reliability analysis of coherent multistate systems, *IEEE Transactions on Reliability*, **44** (4), 683-688.

Yamada, S. and Nishiwaki, I.M., (1995), Optimal allocation policies for testing-resource based on a software reliability growth model, *Mathematical and Computer Modelling*, **22** (10-12), 295-301.

Yamada, S., Ohba, M. and Osaki, S. (1984), S-shaped software reliability growth models and their applications, *IEEE Transactions on Reliability*, **R-33** (4), 289-292.

Yamada, S. and Ohtera, H. (1990), Software reliability growth models for testing-effort control, *European Journal Operational Research*, **46**, 343-349.

Yamada, S. and Osaki, S. (1985), Software reliability growth modeling: models and applications, *IEEE Transactions on Software Engineering*, **11**, 1431-1437.

Yamada, S., Tamura, Y. and Kimura, M. (2000), A software reliability growth model for a distributed development environment, *Electronics and Communications in Japan Part III*, **83** (12): 1-8.

Yang, B. and Xie, M. (2000), A study of operational and testing reliability in software reliability analysis, *Reliability Engineering & System Safety*, **70**, 323-329.

Yang, B. and Xie, M. (2001), Optimal testing-time allocation for modular systems, *International Journal of Quality and Reliability Management*, **18** (8), 854-863.

Yeh, W.C. (2003), An evaluation of the multi-state node networks reliability using the traditional binary-state networks reliability algorithm, *Reliability Engineering & System Safety*, **81** (1), 1-7.

Zequeira, R.I. (2000), A model for Bayesian software reliability analysis, *Quality and Reliability Engineering International*, **16** (3), 187-193.

Zhang, T.L. and Horigome, M. (2001), Availability and reliability of system with dependent components and time-varying failure and repair rates, *IEEE Transactions on Reliability*, **50**, 151-158.

Zhang, X.M. and Pham, H. (2002), Predicting operational software availability and its applications to telecommunication systems, *International Journal of Systems Science*, **33** (11), 923-930.

Zhao, R. and Liu, B. (2003), Stochastic programming models for general redundancy-optimization problems, *IEEE Transactions on Reliability*, **52** (2), 181-191.

Zheng, S.H. (2002), Dynamic release policies for software systems with a reliability constraint, *IIE Transactions*, **34**, (3), 253-262.

Index

MIX
Papier aus verantwortungsvollen Quellen
Paper from responsible sources
FSC® C105338
www.fsc.org

If you have any concerns about our products,
you can contact us on
ProductSafety@springernature.com

In case Publisher is established outside the EU,
the EU authorized representative is:
Springer Nature Customer Service Center GmbH
Europaplatz 3, 69115 Heidelberg, Germany

Printed by Libri Plureos GmbH
in Hamburg, Germany